MALLORCA

The making of the landscape

MALLORCA

The making of the landscape

Richard J. Buswell

DUNEDIN
EDINBURGH ◆ LONDON

First Published in 2013 by
Dunedin Academic Press Ltd
Head Office:
Hudson House, 8 Albany Street,
Edinburgh EH1 3QB

London Office:
The Towers, 54 Vartry Road,
London N15 6PU

ISBN 978-1-78046-010-9
© 2013 Richard J. Buswell

British Library Cataloguing in Publication Data
A catalogue record for this book is available from the British Library

Typeset by Makar Publishing Production
Printed and bound by CPI Group (UK) Ltd, Croydon, CR0 4YY

Contents

List of Illustrations

List of Illustrations

Acknowledgements

This is not a book based on original research in the sense of a dependence upon primary resources, but it has been constructed by drawing upon work largely written in Catalan and Castilian and interpreted for the benefit of British and other non-'Spanish' readers. Like any such book, this means reliance upon the work of many scholars to whom I am in debt; I trust that they will find that I have not misrepresented or interpreted their works, recorded in the bibliography, too clumsily.

Such a book relies upon good libraries, and in this I have been most fortunate. The Biblioteca Bartolomé March near the cathedral in Palma is the best repository of secondary material on Mallorca and contains a great many primary sources too. No one writing about Mallorca can avoid using its facilities. I owe a great debt to its director, Snr. Fausto Roldán Sierra, and his attentive staff, who coped patiently with my halting Spanish. Without this library and its excellent service and working conditions, this book would not have been possible. Biblioteca March is as libraries ought to be: spacious, quiet and not dominated by the ubiquitous computer. I would like to thank Dr Gonçal López Nadal for introducing me to its charms.

Nonetheless, computers are a vital part of the armoury of any writer today. Those serving the catalogue of the University Library at Cardiff were essential for much of the contextual research that lies behind this book. The stock and the staff of the Arts and Social Sciences Library, in particular, proved most helpful to me.

Mallorca is fortunate in having a fine system of municipal libraries, and two of them provided material and good working conditions. That at Manacor is housed in one of the ranges of the sixteenth century convent of Sant Vicenç Ferrer. Where else can you work looking out over late medieval cloisters with the scent of orange blossom wafting up from below and the sound of a cock crowing in the near distance? That at Artà is located in a splendid eighteenth century house (Na Batlessa) given to the town, and many a morning I sat working in its cool reading room listening to the quiet chatter of mothers and their children in the sunny street outside.

Two Mallorcan colleagues require special thanks: Dr. Gonçal López Nadal of the Institute of Economic History at the University of the Balearic Islands (UIB) and Professor Isobel Moll i Blanes, Professor Emeritus of Contemporary History of the same institution. Their conversation and erudition have guided me to sources I would never have found otherwise. Both have read early drafts of parts of this book, helping to ensure that I avoided many infelicitous mistakes in my understanding of Mallorcan history. Dr López Nadal was especially helpful with a revision of Chapter 8. More to the point, both have become good friends.

My former colleague and good friend, Dr Michael Barke, Reader in Human Geography at Northumbria University, also gave up valuable time to read an early draft, and his perceptive comments have made this a better book.

Dr Volker Stalmann provided many stimulating conversations on the origins of talayots and navetes and provided Fig. 8. 2.

The maps and diagrams have been expertly drawn by Alun Rogers of the School of Earth and Ocean Sciences, Cardiff University.

Thanks to Yvonne Owens for help with word processing.

I am indebted to Francesc Carulla Riera of Estop S. A. for permission to use the aerial photos in this book and to the doyen of Mallorcan photography, Andreu Muntaner Darder, for permission to use material from his unrivalled archive.

In drawing upon so many published sources there is always a danger that proper acknowledgement in the references and endnotes may have been over-looked. However, I believe all the material used in this book falls within the sphere of 'fair dealing' and the usual academic protocols. Should any writer or publisher feel otherwise I, and the publishers, would be pleased to hear from them.

Finally, my partner Hawys Pritchard, herself an accomplished Hispanist and translator, has given me much encouragement in writing this book. Her affectionate tolerance of my frequent absences in various libraries and in the field when I should really have been with her are much appreciated.

No fellowships, scholarships, bursaries or research grants have been consumed in the research and writing of this book.

In the end the usual caveats and riders apply, namely, any errors are mine alone.

Palmasol, Mallorca and Cowbridge, Wales
March 1st 2012
Dia de les Illes Balears and Dydd Gŵyl Dewi

Preface

This book is the outcome of visiting and living in Mallorca on and off for more than twenty years. Early visits were as a tourist, one of about one and a half million Britons who visited the island annually in the late 1980s. Initially stays were in small hotels or rented villas, but with buying a house longer visits could be contemplated. In the mid-1980s I undertook fieldwork and some more extensive research into the tourist industry. This also enabled me to make contact with Mallorcan geographers and those involved commercially and politically in that industry (Buswell, 1996). More recently this interest in the island's tourism has resulted in a book (Buswell, 2011). In addition, my interest developed in the urban and regional development processes in the island and in urban conservation, both areas in which I had undertaken research in England, and subjects which the Govern Balear and the Ajuntament of Palma were beginning to develop.

Over the last ten years and with the benefit of longer periods staying in Mallorca, I began to pursue a longstanding interest in the history of the island's landscape, but soon realised that my knowledge of the processes at work in Mallorca was woefully scant, except to realise that they were not like those operating in England and Wales, with which I was more familiar. What followed was a fairly intense period in the field and in the library, a period of discovery.

I also realised that my situation was probably not very dissimilar to that of thousands of others who visit the island each year, more and more of them not members of the usual package holiday market of the high season. They too must have asked questions about the appearance of the landscape: how it got to be as it is, what kinds of pressures had brought about so many changes; had the high rise hotel and the holiday apartment obliterated previous features? Once they began to travel outside the confines of the beach resort where, regrettably, 'all-inclusive' packages corral the holiday-maker even more, they would soon experience the diversity of life and landscape of Mallorca and, if off-season, appreciate that there is a much older and culturally richer Mallorca beneath the veneer of modern coastal tourism. Although there are many histories of the island, few of them seem to deal with these landscape-related questions.

The overseas academic community was also increasing its interest in the island from a whole range of viewpoints. Historians began to familiarise themselves with the archival resources of Mallorca, geographers began to study landforms and environment, sociologists and anthropologists became interested in rural society, archaeologists were tempted by the debates around Bronze Age culture, and that very diverse group of academics whose subject was the tourism industry found one of the most accessible laboratories in the Mediterranean. Many university departments began to organise field courses in geography, geology and biology on the island; the Erasmus programme for student exchange brought more and more undergraduates from across Europe to spend part of their education there.

Mallorcan and Balearic academics' own concerns for their island, its environment and its society, naturally go back much further with pioneering work by geographers such as Barceló, Rosselló and Salvà and historians such as Quadrado, Santamaria and Manera – and all of us owe a debt, of course, to Archduke Luis Salvador's labours published in the 1880s. But it was not really until the post-Franco era and the founding of the present-day University of the Balearic Islands (UIB) that research and publication began to take off, almost exponentially. Over the last thirty years their output of writing about Mallorca has been prodigious in all the fields that relate to this book's content, as the numerous references and the bibliography demonstrate.

It is these works and their predecessors in Catalan and Castilian that I have mostly drawn upon to construct the necessary synthesis in order to make sense of 'the making of the landscape', an aspect of the interface between geography, history, archaeology and ecology that local toilers in the field may have been less concerned to develop; they have had other priorities. In the Acknowledgements I set out my considerable debt to Mallorcan and Spanish academics and to the increasing number of British and other European workers who, like me, have discovered the delights of Mallorca and its people as objects of study. I hope that the outcome of this unabashed borrowing, transliteration, nay, occasional plundering of the work of others will meet with their approval. It seemed pointless when I set out on this task to 'reinvent the wheel' that they had so wonderfully constructed. I suspect that, with very notable exceptions, much of the material I have consulted will be largely unfamiliar to non-Spanish researchers. What I have tried to do is reassemble or synthesise it in a way that will be fairly familiar to British and German readers with an interest in landscape history, but which may be less familiar as an exercise to local students. However, there is a small number of classic Mallorcan texts by geographers that has made a notable contribution to the theme – and content – of this book; most importantly, Vicente Rosselló Verger's magisterial work on the south and south east of the island (Rosselló Verger, 1964) and Miguel Ferrer Flórez's two volumes on the Tramuntana (Ferrer Flórez, 1974), both of which followed in the traditions of French regional studies in which history and geography are closely interrelated. In French, Jean Bisson's *La Terre et l'homme aux Iles Baléares* (1977) is another such classic. In the systematic fields Bartomeu Barceló i Pons' work on population (Barceló Pons, 1970), Albert Quintana Peñuela's on Mallorca's urban system (Quintana Peñuela, 1979) and the work of many geographers and others at U. I. B. in tourism studies, including Salvà Tomàs, Picornell Bauçà and Seguí Pons, many of whose works are listed in the bibliography, have helped condition my thinking on the development of Mallorca's landscape. Finally, no one should underestimate the value of the *Gran Enciclopèdia de Mallorca* as an initial source of reference. Published in a series of volumes from the late 1980s onwards, it has contributions from many of Mallorca's finest scholars.

Despite all these influences, the book remains the work of a geographer. We are a breed who, in Jay Appleton's words, 'have had the experience of trespassing in other people's territory without apology or even an honest blush. . .' (Appleton, 1975: 5) but what we do bring to bear is an ability to assemble the findings of workers in many systematic fields and draw them together in new and interesting ways; in this case, in the service of understanding the origins of Mallorca's landscape and in offering an entrée into its interpretation.

Naturally, the interpretation presented here is that of a white, Anglo-Saxon visitor, no matter how much time he has spent on the island, no matter how many books and journals he has read nor the number of local workers he has consulted. If the concept of landscape history is less well known and practised in Mallorca, so its writing by an English geographer – and a retired one at that – is the product of his perceptions. Indeed, this is one of the contradictions in this field, which the writer cannot escape. If a Mallorcan or a Spaniard were writing this book, no doubt it would have a different viewpoint. The history of the island's landscape resonates quite differently according to one's own cultural upbringing, a topic briefly referred to in the Introduction and the final chapter.

By definition, a work like this, which attempts to cover at least four and a half thousand years, has to be broad in content if narrow in concept. This has meant that detail on certain aspects of the island's landscape history have had to be sacrificed to the need to provide a comprehensive cover. Those with more knowledge of Mallorca will no doubt point to faults of omission, but for the many readers less familiar with the island, this book should provide a useful introduction to the topic and point the way to further research. In particular I have been obliged to omit some of the smaller, more intimate elements that go to make up a landscape. I have kept to the wider canvas of a chronology that deals with change over time.

The city of Palma – Ciutat – presents a particular challenge to anyone writing about Mallorca's landscape. In the medieval and early modern period and again in today's more metropolitan era and at a time of counter-urbanisation, this city has loomed large in the island's history. In landscape terms it really demands a lengthy chapter to itself, but in fact, so rich and diverse is its townscape that only a book would suffice. Readers with a bent for urban history will have to forgive the author if they find his treatment of the city's form and function over two thousand years or more somewhat superficial; it is not for want of interest or competence but for lack of space. However, certain aspects of Palma's historical geography and townscape may be found in most chapters.

Finally, the reader will become aware that the author has drawn many of the more detailed examples of processes and patterns of landscape change from the eastern side of the island and from the municipality of Manacor especially. His excuse is one of familiarity and local experience of this particular corner of Mallorca; it is his 'back yard'. Geographers will be familiar with the philosophical niceties bound up in this constraint.

A note on languages
Mallorca has become an island of many languages. Spanish (Castilian) and Catalan are the official ones, but the language spoken by many of the inhabitants is Mallorquin, a distinct and proper dialect of Catalan. In the tourist areas English and increasingly German are spoken by many, but one hears many others, including the Latin American version of Castilian and Magrahbian dialects of Arabic, thanks to recent immigration. For this book it seemed correct to quote and use, where most appropriate, the written form of what many Mallorcans believe to be their first language, Catalan.

A note on Mallorcan measurements
Although the standard European metric measurements were adopted by Spain, and by supposition, by Mallorca in the nineteenth century, the sources and the literature are full of references to much older measures. Indeed, in the rural areas many of these are still in everyday use. For more details see: Casanova y Todolí and Lopez Bonet, 1986.

- A *jovat* or *jovada* (Castilian *yugada*) was said to be the area of land that could be ploughed in one day with a pair of oxen. 1 jovada = 11. 36 ha approx.
- A *quarterada* = 0. 7103 ha and its subdivisions, the *quarta* (1/4 of a quarterada) = 0. 1783 ha (1783 m²) and the *horta*, (1/16th of a quarterada) = 0. 0444 ha (444 m²).
- A *palmo* = 19. 55 cm.
- A *cana* = 8 *palmos* = 156. 4 cm.
- A *vara* = ½ *cana* or 4 *palmos* =78. 2 cm in Mallorca but all provinces had their own *vara*; a national standard was adopted in 1801 = 86 cm (a 'rod' or a 'pole').
- A *destre* = 4. 214 m = 12 *palms de destre*. It can also mean an areal measure, i. e. a *destre cuadrado* = 17. 75 m².
- A *legua real* = 6687 m.
- A *legua maritima* = 5573 m.
- A *libra* (pound weight of 16 oz) = 0. 407 kg.
- A *quartan* = 4. 15 litres (almost a UK gallon).
- A pre-metric currency existed similar to the old British money of pounds, shillings and pence. The largest monetary unit was called *Lliura* (pound) which was based on the weight of silver equivalent to 327 gr. The *Lliura* was divided into 20 *Sous* which in turn had a value of 12 *Diners* each. The metric monetary system was introduced in Spain in 1869 when the *Peseta* replaced

Lliures, Sous and *Diners* in Mallorca. Since 2002 Mallorca, as part of Spain, has been a member of the Eurozone.

- *Morabatí* (Castilian: *maravedi*) originally a coinage introduced by Alfonso VIII of Castile to enable trade with the Muslims whose *dinar almorávide* was judged equivalent. Its value varied but usually 1 *morabatí* = 8 *sou* or 96 *diners*. Widely used as a measure in taxation.
- *Focs* (literally *fires*) hearths in houses, used as a measure of wealth and as a source for calculating taxation.

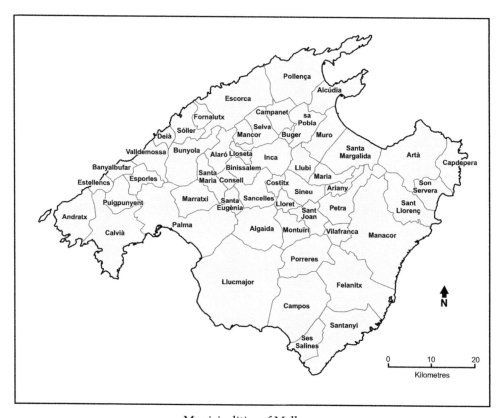

Municipalities of Mallorca

1

Introduction:
Mallorca and landscape history

Indeed, at any point in history a landscape always consists largely of its past.
(Fred Aalen, 2001)

Mediterranean landscapes are human artefacts in which complex cultural
histories are firmly embedded. These landscapes should be interpreted
as manifestations of historically specific identities shaped by different
human societies over many millennia according to deep-seated cultural
principles ... it is clear that these landscapes are also shaped by the practical
constraints of climate, geography and geomorphology, and the biology of
plants and animals. Technological practices and developments ... are critical
for understanding how human societies exploited these landscapes.
(Lin Foxhall, 2003)

Arrival

*As the tourist's no-frills flight from Berlin or Manchester loses height over the
Mediterranean after crossing Catalunya's shore and turns right over a pair of large
bays, the view from the aircraft window shows a range of quite high mountains to the
right, a fairly flat plain with some low hills straight ahead and a line of higher, rising
land to the left. The mountainsides, whose tops are clearly sloped with bare limestone,
are clothed in dark trees. Closer inspection reveals a network of ochre, brown, almost
orange fields interspersed with patches of woodland. These fields seem to be patterned
by shortish trees laid out in serried ranks. Many are surrounded by rectilinear low stone
walls; others appear to have no solid divisions at all. Much of what can be seen has a
geometric appearance. Over the centre of the island quite large towns can be made out,
and what looks like a high speed motorway snakes its way south-westwards, another
northwards. The land is dotted with numerous isolated farmhouses with sloping tile
roofs. Other buildings appear to have swimming pools. Below is a wide, flat plain nailed
down by brightly coloured windmills, few of which seem to turn in the breeze. Suddenly
below is a large city with sprawling suburbs, endless blocks of apartments and industrial
estates. To the keen observer, a tightly-packed network of narrow streets can be seen at its
core, dominated by what must be a massive Gothic cathedral. The plane moves out over
another large bay, banks left and now, set out to left and right, almost from one end of
the low horizon to the other, is a continuous built-up area along the sea's edge. Dipping
low it comes in to land over acres of grey concrete almost surrounded by hangars and*

warehouses, and, curiously, what appears to be field upon field of a strange crop. The
farmers here grow cars!

Our visitor has passed over a complex scene of old and some patently new elements. Where did they come from? What are their ages? Many are clearly the product of the modern era, but some appear ancient. Are both the product of the so-called 'timeless Mediterranean'? Are they unique or are they like similar features in urban, industrial Europe further north? Unravelling this complexity will take a combination of the skills of the geographer and the historian even for what is, after all, a small dot in the sea, an island about twice the size of the old county of Glamorgan in Wales, but a 'dot' that in 2012 received over nine million visitors, about 2500 for every square kilometre if they all came at once. Luckily they don't, but as we shall see, there is still a remarkable seasonality that results in very high densities in July and August. Our visitor is suddenly a key figure in this island landscape, an agent of its continuing change.

It is clear from this parody that one force above all that has shaped today's island landscape has been mass tourism. Over the last fifty years it has had more effect than any other in altering four and a half thousand years of relatively slow change. Although this book will set out the ways in which economy and society have wrought new features and how, over much, much longer periods, nature has constantly refashioned the island's surface, the impact of the aeroplane, the hotel, the apartment, the motor car and nine million visitors a year have transformed a largely agricultural environment and a trading city into a modern built-up land-scape that increasingly resembles parts of north-west Europe. Indeed, as second home ownership and more permanent immigration increases so the population, language and culture take on more than a tinge of Düsseldorf, Manchester and Rotterdam, as well as Madrid and Barcelona and Marrakesh.

This presents the landscape historian with a major problem. The rates and scale of recent change historically have been so great that previous elements have simply not survived in many, especially coastal, areas. We are often not dealing with relict features – survivals – but with completely new landscapes, so that only a future generation of archaeological digs will be able to interpret what came before 1960.

There are some constants, however, that can be isolated to help us build up a more systematic analysis.

First is the Mediterranean setting, discussed in more detail in the first two chapters. Clearly it has been the Mediterranean and its physical geography that lies at the heart of the success of the tourism industry, especially the school pupil's facile reduction of its complex climatic and meteorological characteristic to 'warm wet winters and hot dry summers'. The geology of much of the Western Mediterranean, with its sedimentary rocks and alpine-type folding, has given coves and headlands, beach material and mountains and hills for scenic variation.

But the Mediterranean is a sea, and Mallorca is an island, and long before the first tourist sailed into Palma's harbour these clichés, together with the actual position of the island, 120 km from the Spanish Peninsula, 200 km from North Africa and 150 km from Europe to the north, meant that though 'isolated', in one sense Mallorca has been at the heart of the western basin, in another, a veritable crossroads (see Fig. 2. 1). Even as early as AD 1200 Mallorcan traders were operating all along the North African coast and the coasts of Languedoc and northern Italy. By 1350 their ships were as far afield as London and Antwerp.

This has meant that Mallorca has been on the receiving end of numerous settlers and colonisers, and not only human ones. If isolation meant that ecologically the island developed small and sometimes unique populations, its geographical location meant it was accessible to many water-borne, and later air-borne, cultures and species. Its peoples, especially, have origins in the distant past in the Levant, the Maghreb, the Italian peninsula and Catalunya-Aragon, and in the twentieth century from all over Northern and Western Europe and particularly from Andalucía, Murcia and Valencia. In the early twenty-first century the growing numbers of immigrants from North Africa and Latin America under Spain's fairly liberal policies will in time surely have an impact too. 'Invasion and succession' is a constant theme in biology; it is a characteristic, too, of prehistoric and historic human settlement, with each 'wave' bringing new landscape elements and processes and helping to obliterate, modify or mask existing ones.

Much of our analysis, then, will be concerned to demonstrate the effects successive peoples and cultures have had in shaping what we see today in Mallorca. But Mallorca has its own take on these forces, and developed indigenous and autochthonous features too. For example, the *talayots* of the late Bronze Age are not like settlements of the same age elsewhere in the Mediterranean, indeed in Mallorca they are even different from those in nearby Menorca. Industrialisation in the nineteenth century certainly did not follow the 'English' model pursued in so much of the rest of Europe. Until the 1990s these influences were largely regional, that is, largely west European, but today Mallorca is as much part of a globalising world as the rest of Spain, as inward and outgoing investment reach beyond the shores of the Mediterranean. What we are saying is that, despite the influence of geographically wide forces, the landscape of Mallorca has an interpretation of its own, a local distinctiveness. The trick is unravelling the degree to which outside forces have been at work, the way purely local forces have operated, and how the resulting landscape is an amalgam of all of these.

One last introductory point is to deal with the place-to-place variations on the island itself, its own spatial organisation. Although, even by Mediterranean standards, Mallorca is a small island (3640 km²), there are considerable variations in the geography. Many of these can be attributed to the physical structures: the mountains in the north and west, the Levant (eastern) hills, the Pla (the central plain), the variations in coastal features; but historically there has been, at least since the Roman colonisation, a marked dichotomy between *ciudad/Palma* (the City of Mallorca) and *part forana* (the remainder of the island, essentially the countryside) (Rullan Salamanca, 1998). Certainly since the Muslim invaders developed their Medina Mayurqa in the twelfth century, Mallorca has been dominated economically and culturally by its primate city, Palma. The Catalan word for this is *macrocefalic* – large headed. The struggle between centre and periphery has erupted from time to time, such as in the *germans* (the 'brotherhoods') uprising of the sixteenth century, but generally speaking Mallorca has been synonymous with '*Ciudad de Mallorca*', its hinterland backward and remote, a source of potential agricultural wealth but often a cultural backwater, largely thanks to a neglectful and often absentee aristocracy. Despite attempts to push the economic frontier into the backlands, such as the new town movement following the Ordinances of 1300, despite the agricultural reforms of the eighteenth century with the introduction of tree crops, despite the industrialisation of some towns such as Manacor and Inca in the nineteenth and twentieth centuries, the *part forana* remained, above all, poor. Even today the Palma municipality contains a remarkable 57% of the island's population, and as counter-urbanisation spreads into Calvià, Llucmajor and Marratxí, the city-region of Palma grows ever wealthier, and its landscape and townscapes reflect this with their high-rise blocks, their suburban apartments, industrial estates and the expanding high speed road system.

Earlier we asked the 'who, whom' question of cultural landscape formation over time. As part of his explanation as to why tourists came to Mallorca in the early 1950s, Robert Graves said it certainly was not because of the island's historical attractions. His meagre shortlist included a castle (Bellver) that had never been besieged, a fifteenth century peasants' revolt, an invasion by King James in 1229 and a disastrous attempt at conquest by Republican forces during the Civil War – 'no memorable passage of arms has happened here' (Graves and Hogarth, 1965: 45). In the case of Mallorca, over the last five millennia, the actors range through Bronze Age tribes, Roman army units, Berber clans, medieval kings and lords, small farm co-operators, Fascist centralists and tour operators to technocratic planners and elected politicians at many levels. Underpinning all of these have been individuals, families and small groups: the *roter* (agricultural labourer) clearing the land and making fields in the eighteenth century; the *pagès* (the small farmer) and their families experimenting with new crop rotations, the aristocrat and his eighteenth century olive oil empire, the Roman veteran, the Andalusian bricklayer in the 1960s, the Moroccan dry-stone waller of the 1990s. Most of these individuals are lost to us; only occasionally do significant names surface in the archives: J. R. Bateman and his plan for reclaiming the Albufera, King Jaume II and his plans for new towns in the fourteenth century, Giacomo Paleario and his brother Giorgio – the designers of Palma's Renaissance walls, Eusabio Estado and his plan for the railways and Jaume Ferrer and his revival of viticulture around Binissalem in the 1970s. Landscape making (and destroying) is a 'people process'. In Mallorca many of these have been native islanders, but many have been in-comers ranging from conquerors to modern investors, from colonisers to day-labourers. All have contributed to the landscape of the Mallorca of the twenty-first century.

Landscape and its history

Interest in 'landscape', and in particular the history of landscape, is currently undergoing something of a revival; 'place' and 'geography' are now near the top of the 'cultural studies' agenda. If writers and painters have always retained the landscape as an inspiration and as a subject from the Picturesque and the Romantic periods onwards, then for historians and geographers, after a period of fairly intense concern in Britain in the 1950s and 1960s, interest faded away under the assault of supposedly more objective and scientific approaches to their subjects, and in the case of human geography, their concern in the 1980s for a more determined political and politicised approach. During the last twenty years a very wide spectrum of perspectives and critical stances has been advanced under the heading of 'cultural geography' (Wiley, 2007) and to a lesser extent in landscape archaeology (Bender, 2002).

What has led to the recent revival has been the development of cultural geography on the one hand and landscape archaeology on the other. The former is a subject that for some has its origins in the work of Carl Otwin Sauer and the Californian School of the 1950s. Sauer and his followers largely studied remote and rural landscapes, not those of urban-industrial America. Another American geographer, J. B. Jackson, advocated an approach based on the vernacular aspects of landscape – the everyday, the ordinary, the mundane, the workaday world around us of which we are part – producer and consumer of landscape. In post-war Britain, W. G. Hoskins, being a historian – and essentially a local historian of the countryside and small towns and provincial cities – developed 'an intense sense of temporality', emphasising through a chronological method, 'the depth, richness and complexity granted by sheer cultural age' (Wiley, 2007: 32) – that is, the idea of the deep past, not just the superficial but widespread changes of the late nineteenth and twentieth centuries. For him, landscape was a material entity – something to touch, observe, walk in – not just a picture or its dematerialised symbolic meaning (Wiley, 2007: 54). A tension between the 'past' and the 'present' is always evident in most of his writings; what David Matless has called 'tradition embattled by modernity' (Matless, 1993: 191).

If these three 'founding fathers' of landscape history focused primarily on the material world, the present paradigm is now more influenced by concerns for perception, social construction, values and viewpoints with leanings towards social psychology, sociology, literature and history – of art and society. A recent research topic from the British Arts and Humanities Research Council was also indicative of this approach – chaired by a geographer but with activists drawn from many humanities and social science backgrounds. [1] Landscape archaeology emerged from traditional archaeology, which originally had a focus in buried artefacts, when it became necessary to see and interpret 'finds' in a variety of contexts, including the material, the socio-economic and the cognitive. The exhumed sporadic objects, buildings and settlements were seen to be part of wider spatial systems: the whole landscape had an archaeology. Once traditional archaeology moved out of the pre-historic and into the historic period, and especially when it embraced the medieval and post-medieval, the ground was cleared for this landscape approach. Clearly in a post-modern age there has now emerged a range of models and theories from a wide variety of disciplines that have come together to produce the current diversity of paradigms for the study and analysis of landscape and its history.

The conventional model for the history of landscape amongst many British historians – if not amongst cultural geographers – remains the approach devised by Professor W. G. Hoskins and the school of local history that he founded at the University of Leicester, UK. Besides his own seminal Making of the English Landscape (1955), Hoskins edited a series of volumes largely written by a range of historians and geographers on many, but by no means all, of the English and Welsh counties. In the same vein there followed books on the Scottish landscape, the Dutch landscape, the landscape of the United States, even a series on the 'World's Landscapes' with individual volumes on Wales, Ireland, China and the Soviet Union. Another series with a thematic approach was the 'Archaeology in the Field Series' published by J. M. Dent with works on fields, trees and woodland, and towns. Both Hoskins' and the 'World' series remain incomplete; perhaps the paradigm had run its course; although more recently a new edition of Hoskins' volume on England volume updated by Christopher Taylor (Hoskins, 1988). More recently still, Trevor Rowley has ventured to produce a book on the only just deceased twentieth century (Rowley, 2006). Still in the Hoskinian tradition but with a powerful input from archaeology is Francis Pryor's The Making of the British Landscape (2010). For the Mediterranean region one of the most recent in this long line, but written by a botanist and an archaeologist, is the Making of the Cretan Landscape (Rackham and Moody, 1996) together with Vogiatzakis, Pungetti and Mannion's edited essays Mediterranean Island Landscapes: Natural and Cultural Approaches (2008).

Although some British and Irish geographers such as Dennis Cosgrove, Stephen Daniels, Richard Muir and Fred Aalen used an older flame to light up new directions during the period since the 1970s, progress within the older tradition has perhaps been made by archaeologists and historians including, inter alia, Christopher Taylor, Tom Williamson and Richard Newman. Much more satisfying has been the co-operation between a whole range of landscape disciplines, good examples of which might include English Heritage's eight volume series on England's Landscapes (Cossens, 2006) and the magisterial Atlas of the Irish Rural Landscape (Aalen et al., 1997).

Amongst German geographers their interest in landschaft is of longer stand-ing, whether for rural landscapes or the morphological aspects of the urban scene.

Spanish geographers' early affinity was with their French neighbours' *géographie humaine*, although they have also tended to follow German teachings, certainly in the 1980s; but during the 1990s there was little concern for landscape as they pursued other leads set down by American and British human geographers (Luis Gomez, 1980 and 1984). However, in the early twenty-first century there seems to have been a revival of interest in the links between geography and history amongst some Spanish geographers, perhaps because the landscape in so many parts of their national territory has been changing rapidly, especially from the effects of tourism in coastal zones. [2] Nicolás Ortega Cantero and his colleagues associated with the Instituto del Paisaje de la Fundación de la Duques de Soria, for example, have emphasised the close affiliation between geography and history in Spanish thought and academic organisation and practice. They have re-explored, on the one hand, the influence of the French School of human geography epitomised by the work of Vidal de la Blache, which emphasised the importance of culture in understanding landscape, and on the other, the German School descended from the work of von Humboldt and Karl Ritter, which pursued the role of natural or scientific factors in landscape formation:

> El paisaje, es, para el geográfo moderno, materialidad y formal, pero es también, al tiempo una representación culturalmente ordenada y valorada de esa realidad material y formal ... Naturaleza y cultura se dan en el paisajismo geográfico moderno.
>
> (Landscape is, in modern geography, material and formal but has at the same time an ordered and valued cultural representation that is also material and formal ... Nature and culture go hand in hand in the modern geographical study of landscape.) [Ortega Cantero, 2004: 10]

More recently, Ortega Cantero has written an introduction to a series of essays in *Estudios Geográficos* that builds on this previous work and reassesses the role of the study of landscape in modern geography in Spain (Ortego Cantero, 2004, 2005 and 2010). Two Mallorcan geographers have recently risen to the challenge too (Binimelis Sebastián and Ordinas Garau, 2012).

Throughout all these five or six decades the environmental movement also provided a stimulus as scientists and the general public sought to understand the origins and process that had brought about the landscapes said to be under pressure from developments that were both ecologically degrading and destructive of a fragile heritage.

There has always been a persistent undercurrent of concern for landscape history. Sometimes located in odd corners of academia and amongst the general but educated public, research into the origins of the socially constructed surface of the earth has been going on, usually on fairly local scales. This is connected with a broader concern for 'the past', perhaps even with a nostalgic belief that times past were better than times present. As younger people seem to know less and less about history – certainly of more distant periods – so an older generation seems keen to know more about their origins and the places they inhabit. It seems to be part of a movement to discover and be reassured about rootedness, a sense of belonging, whether to family or place. Maybe this is a reaction to the increasing pace of life and 150 years of social and geographical mobility. A parallel may be found in the active survival of the study of urban morphology, where Professor Jeremy Whitehand and his colleagues at Birmingham University kept the lamp burning that was lit many years ago by Professor M. R. G. Conzen. In the case of the history and making of landscapes, all these acknowledge a debt to one of the 'founding fathers', W. G. Hoskins.

In the more 'cultural' approach of recent years some geographers have again been at the forefront, not dismissive of the historical/geographical approach by any means, but propounding one that seeks to shift understanding and interpretation away from a largely chronological, developmental approach – in which the changing relation between society and its environment has something of a whiff of nineteenth century determinism about it (despite the usual protestations) – and towards a more culturally bound view in which social class, gender, economic position, ethnic and ideological values and psychological perception are used to interpret landscapes of previous and present eras. Many archaeologists have moved in a similar direction.

If landscapes are cultural artefacts, the outcome of social processes, then it is reasonable to ask who the men and women were who fashioned them. Social, economic and political 'forces' are all very well, but behind these were real people, at one level individuals but at another social groups, classes or communities. Are landscapes the product of kingly edict, of local aristocrats, of tribes or clans, of the elected or non-elected State or of communal peasants? There is no 'progress' implied in this list, from a hierarchy to a modern parliament, no movement necessarily from an autocratic to a democratic landscape. Nonetheless, many of the large-scale technological changes or innovations that altered Mallorca's landscape, such as terracing and irrigation, were often a product of the power of a central authority over small-scale subsistence production. One may even cite the example of the built environment of the touristic landscape of the 1960s of high rise hotels, apartments, promenades, etc. as products of Fascist planning and architecture imposed in an era of economic isolation and the need for foreign exchange. The influence of the small farmer and the villager is just as evident to those who look closely, but it has been more limited, to the *horta* and the smallholding. As this story of Mallorca's landscape unfolds we shall have to be conscious of the balance of political power in its production.

Landscape has inevitably got mixed up with the murky world of the heritage industry, whether the state sponsored or the private sector versions. Some of this has given rationale and legitimacy to conservation, but it has also been hijacked for the commodification and selling of landscape, often based on poorly researched history and geography, relying more on myth and legend for the development of visitor attractions, theme parks and their ilk. The '*story*' of Mallorca's landscape history is, then, often different from its *history*, a view of *Mallorcan-ness* devised for commercial as well as political purposes. This moves us to ask, 'who owns the history of this landscape?' Is it that share of Mallorca's population whose family roots go back many centuries, tracing their origins from amongst King Jaume I's conquering army? Or is it the Andalusians' and Murcians' who were attracted here in the 1960s by the building boom as Mallorca's tourism industry 'took off'? Or is it the families' of the Moroccan and Latin American immigrants who have arrived in the last decade? Does it belong to the second-homes owners from Germany and Britain? Some will seek to make stronger claims to this history than others. This author has chosen to base his interpretation on a conventional Mallorcan one set in the context of an emerging Spanish State. Such a view may seem obvious at first, but we have already sowed some seeds of doubt as to who the Mallorcans were and are. To date, the political power behind Mallorcan history has been that of the educated and native born who would appear to have determined what to include, but new peoples and cultures in a rapidly changing society may well have other priorities. American scholars and others have already undertaken detailed studies of medieval Jewry in Mallorca. In future the academy can look forward to more archaeology of the Muslim period, to studies of German settlement in rural Mallorca in the period from 1960 and of the role of Magrhebian stone wall builders in the last two decades, for example. All of these cultures have shaped and are shaping the island's landscape, and as such they and their heirs are certain to lay claim to this history. It may be that one of the functions of a more inclusive history – in which landscape history can play a part – is to help forge a new identity for the island's past. In Britain, elements of this approach were explored by Ralph Samuel (Samuel, 1994: 139–168). In Spain, Ortega Cantero and his colleagues certainly see landscape as that which is seen, the perceived scene and, linked to the interpretation of landscape by artists, that helps to create a sense of national identity, of 'Spanishness' (Nogué, 2005: 149). Clearly a notion that can be extended to regional identity, perhaps harking back to the idea of the *pays* explored in French geography of more than a hundred years ago (Gallois, 1908). If this political function can be added to the study of Mallorca's landscape, then also added to it must be an economic one, particularly as those responsible for the development of the tourism industry seek to harness landscape history and its interpretation as a resource for visitors, an alternative to the beach, the shopping mall and the conference centre.

Thus the paradigms shift, but older views and methods don't disappear altogether. There is always someone, perhaps retaining the best of the old ways but pushing them forward in new directions and into new areas. After all, no-one can say that the historical–geographical way of seeing things was 'wrong', any more than the contemporary 'cultural' view is 'right'. Hoskins and those who have followed in his footsteps are still to be found on library shelves, and easily traced in the second-hand bookshops and the Internet. Because of the detailed research into primary and secondary sources they undertook, their works can never be accused of antiquarianism or myth-making.

This history of Mallorca's landscape is certainly empiricist, rooted in objects and events that are tangible and known, seen through encounter and observation – 'landscape is a milieu of solid fact rather than abstract theory', but again, unlike Sauer and Hoskins, this history is not devoted primarily to the rural. After all, today's Mallorca is largely the product of urbanisation brought about by tourism, not perhaps in straightforward land-use percentage terms, but experientially. In addition, one might also say that even the Mallorcan countryside is in so many respects the product of urban forces emanating from one dominant, primate city (Palma) over at least 1000 years: the ruling aristocracy and nobility spent its time in their palatial town houses and later, when their absence from their landholdings resulted in its demise, many of their estates were taken over by city-based professionals – lawyers, notaries, doctors, military officers, merchants and businessmen.

Here will be found aspects of J. B. Jackson's vernacular approach, and like him we shall be at pains to point to the fact that the many apparent similarities in landscape features across the island are just that – apparent – and that Mallorca's stone walls, talayots, rectilinear features, town plans, rural house types, tourism urbanisations and hotel and apartment architecture – when examined in detail in particular locations – soon appear distinctive and different. While our stance is certainly material, it is not devoted to material objects alone, but *much more towards the economic and social historical context in which they are set*. There is, without doubt, an interpretive, gaze-filtered account to be written of the landscape of Mallorca, but this book is not it. Nor is our purpose to write the first landscape history of Mallorca (some parts of which have already been done in Castilian and Catalan) but rather to produce a version of the island's landscape history in English, drawing upon local sources and inevitably filtering them through one man's experiences, eyes, perceptions and interpretations. Hence, this is not a book destined for locals (unless they are anxious to read of an outsider's view) but a book for foreigners, not in the usual tradition of travel writing but more in the footsteps of scientific writing as exemplified by the massive work of Archduke Luis Salvador – a foreigner, a German from Italy, at home in many parts of the Mediterranean (Archduke Ludwig Salvador, 1888). Unlike Ortega Cantero and his colleagues, this author has not adopted some kind of Franco-German compromise in his approach, but has chosen to take a view based on the changing economic and social organisation of the island over time and geographically expressed.

In the Preface we drew attention to an 'archive', but in the case of an outsider writing of Mallorca it seemed more logical to use the wealth of secondary material that is available to bring to a new audience a study of the island's landscape history. Another source of information that complements the archive and the written record is the land itself. Writers on the theme of landscape construction have always been keen fieldworkers. Not for them the ease of the armchair or the library alone. The distinguished economic historian of some generations ago, Sir John Clapham, described historical geography, perhaps apocryphally, as 'economic history with stout boots on'. Subsequent authors have followed where he and generations of geographers trod. Writing in the 1970s Jay Appleton, then of Hull University, noted at that time, 'Fieldwork became the slogan of the dedicated; muddy boots the symbol of authority' (Appleton, 1975: 6). More recently Simon Schama recalled, 'Historians are supposed to reach the past always through texts, occasionally through images that are safely caught in the bell-jar of academic convention: look but don't touch. But one of my best loved teachers . . . always insisted on directly experiencing 'a sense of place', of using 'the archives of the feet' (Schama, 1996:24). Richard Muir, with his notion of *reading* the landscape, recommended a similar on-the-ground approach (Muir, 1984).

Landscape, then, is a product of social and economic processes and the cultural values of a particular time; it is, to use a dated phrase, above all 'man-made'. Practically nowhere in Western Europe is there genuine 'wilderness' where few have ventured. Indeed, in most of Europe the contemporary landscape is the product of thousands of years of change and adaptation. The physical structures have been worked and re-worked many times over, frontiers of settlement and occupation have advanced and retreated both laterally and vertically according to the 'value' given to land at various times. Even quite substantial towns have come and gone. Within the cities land-use values have constantly shifted as urban economies have been restructured. But each era has left some mark, some imprint, however faint, to survive through as a relict feature to the present day. This is what has been called the *palimpsest* approach to landscape. That which we see today is the product of human and physical activity over long time periods. It is a term derived from the study of ancient manuscripts, where the vellum or paper often proved to be a more valuable resource than the writing upon it, so that early writings were erased and the medium recycled. The original text would remain partially – and often enticingly – visible. Take any fragment of land, and it might contain a Neolithic earthwork, the site of a Roman villa, medieval field patterns, perhaps an example of early industrialisation based on mining, water power and charcoal – all under pressure from a modernising and mechanised agriculture with immense power to level, uproot and deep plough away some of these elements from previous ages, creating a fresh scene. A town may contain its Roman and medieval origins in its street plan and plots of land, but its buildings may be of the nineteenth and twentieth centuries; eighteenth century parks and gardens may become frozen relicts in the landscape, but put to modern recreational uses –'fragments that we can see but which are not an integral part of the current order' (Emery, 1969: xiv). Or in Braudel's words 'the scars of ancient wounds'(Braudel, 1988: 31).

The palimpsest, then, is a useful metaphor, perhaps even a definition of landscape for our purposes, but it is not a methodology.

There are two ways of ordering an understanding of landscape change over time: one is a chronological approach, a time-line from the earliest occupation through to the present day, using the fairly conventional but never wholly satis-factory periods of prehistory and history – the late Bronze Age, the Roman, the medieval, etc. A second way is to use themes, tracing their development and degree of survival over time to the present day – rural settlement, agriculture, mining landscapes, urban growth, etc. Neither is ideal, as elements of both should really be drawn upon continuously. However, any text has to be readable, requiring a degree of logic in the presentation of 'facts' and 'ideas'. A brief examination of some previous landscape histories reveals that other authors have struggled with this dilemma. It seems a compromise has to be decided upon if neither approach is to be adopted solely.

Another difficulty is 'place', or in the language of human geography, 'areal differentiation', 'spatial organisation' and 'regional and local variations'. Whatever processes we succeed in isolating in the formation of landscape, few operate con-sistently through geographical space or historical time. For example, immigrants do not settle regularly in terms of location when they come to a new country. Their numbers and density varies from place to place and so will, therefore, their impact on landscape. And time is a continuum, so that as one 'age' succeeds another the characteristics of the first rarely suddenly die out to be rapidly replaced by others. Often cultures and economies from quite different periods survive side by side: an 'old' landscape and the artefacts associated with it survive alongside a 'new' landscape, often for generations. It will be important to be able to show how these older and newer elements are related in geographical space as the history of Mallorca unfolds from its prehistoric origins through to the landscapes of the tourism industry of the twenty-first century.

In this sense it is not 'backward looking' or nostalgic to use a Hoskinian or Sauerian methodology; it can be used as a means of creating a base for others to work from, using perhaps these less modish approaches as a means of lighting a fuse, setting the scene for the study of the ongoing dynamic of contemporary landscape formation and understanding, looking forward to the next layer of the palimpsest. Any author – including this one – is the product of his or her age, has a set of personal values and approaches to it and, while cognizant of contemporary approaches, writes best about what he knows using the tools he has mastered. The reading public, whether general or academic, will judge the outcomes.

2

Mallorca and the Mediterranean

The Mediterranean is not merely geography. (Metvejevic, 1999)

Context is very important in the study of a small island like Mallorca. It is not a country in its own right, although at times it operated as an independent unit to all intents and purpose before the creation of 'nation states' in the sixteenth and seventeenth centuries. We shall show that even before then, and indeed, from almost its very emergence as a home of mankind, it was and remains a place well connected with a wider world. It became fashionable for earlier generations of Mallorcans to blame their relative economic backwardness on a perceived isolation; the island status of the place was said to impose economic costs (Alenya Fuster, 1984: 189–206). Contemporary historians and economists are now at pains to point to Mallorca's relative wealth, regardless of geographical position; for example, Carles Manera's introduction to his economic history is entitled *Unas islas siempre abiertas al exterior* (Islands always open to the outside (world), Manera, 2006). Similarly, the factors and forces that helped shape its landscapes often originated from outside its own territory, often the product of the cultures that invaded and controlled the island that brought with them values whose origins lay elsewhere, ranging in time, for example, from feudalism to mass tourism. At times it is difficult for the outside observer to distinguish the purely local from the imported. Nearly all the major political powers in the Mediterranean basin since late Classical times have been continentally based: the Roman Empire in Italy, Byzantium in Asia Minor, France, Aragon-Catalunya, etc. There are few examples of island-based ones. The Kingdom of Mallorca, despite its possessions in the Peninsula, in Provence and Sardinia, was short-lived, lasting less than seventy years (1276–1344). It was soon absorbed back into the Aragonese Empire from whence it sprang. Few islands have survived separately long enough from their more powerful geographical neighbours so that dominant cultural influences have nearly always come from nearby land masses, even if from different ones at different times; only in a few have autochthonous forces been strong enough to survive for long. To use a phrase from ecology, 'invasion and succession' as a theme, if carefully used, can be applied here. Nearly all the landscape changes we shall be examining seem to have originated from outwith the island. Notwithstanding, we can detect two processes at work within the Mediterranean: firstly the tension between incoming and already existing forces and secondly, the development of a Mallorcan interpretation and subsequent modification of those influences (see Fig. 2. 1).

For the landscape historian of Mallorca the most important context is neither the political setting nor solely the economic one, but instead the wider cultural setting of the Mediterranean. This presumes that this wider geographical milieu is the *fons et origo* of the majority of the cultural influences; but what is the Mediterranean? In recent years an argument has been developing so that it is increasingly difficult to generalise about it in a way that an earlier generation sought to do. This debate might usefully be summarised – and over simplified – as 'Braudel vs. Horden and Purcell and others', the former being to many the doyen of historians of the Mediterranean, the latter group representing the new and emerging schools of thought. Among the newer writers on the Mediterranean who have extended the debate have been David Abulafia (Abulafia, 2011) and Faruk Tabak (Tabak, 2008).

Braudel's history was much influenced by geography, initially the determinism of the nineteenth century, and later by a more possibilistic view exemplified by the human geographer Vidal de la Blache and by the Annales school of historians led by Lefevre. Braudel sought to show that there was a unity to the Mediterranean largely derived from its physical environment, and from the role of the sea as a connecting agent that permitted the spread of a similar culture. It is difficult to trace the origins of this kind of thinking, but might it not have been the pervasive effect of late Renaissance art, the influence of Classics on education from the Enlightenment and from the impact of nineteenth century science on society? The notions of

Figure 2.1 The Western Mediterranean Basin. The Balearic Islands stand at the crossroads between Europe and Africa.

mare nostrum, ship technology, seaborne empires and trade within that closed sea must have been very powerful for a man who lived part of his early life across the western basin in *algérie française*. There is also, then, the idea that 'mediterranean' is an imperialist concept rather like Said's 'orient'.

Horden and Purcell's view is that rather than heralding a beginning of Mediterranean studies, Braudel's great work represents the end of a paradigm, the 'withering of a scholarly tradition that goes back to the founding fathers of regional geography' (Horden and Purcell, 2006: 39). In their massively detailed study they advance the thesis that the Mediterranean does not have the degree of commonality that Braudel and others saw, but rather, 'history *in* (the Mediterranean) seems more *a propos* than history *of* it' (Horden and Purcell, 2000: 43). This comes from a detailed examination of particular places set in their Mediterranean context, focusing on differences as well as similarities. Their history is 'in large measure a history "close to the soil" – and the sea; a historical equivalent to human geography'. Their approach is ecological, but not in the sense of human beings seen simply as functional organisms, as in an older view of that science, but more in the contemporary paradigm that allows for instability, chaos and disturbance and thereby includes flood, drought, fire, disease and epidemic in its analysis and interpretation which, when translated into human action, involves the concept of economic risk. In addition, they include, like Braudel, the interaction between places, but their concern is less with the long-distance hauls of exotic goods but more with the exchanges of the commodities that sustained everyday life – grain, oil, wine, timber and metals – across what David Abulafia calls an 'enriching' rather than the 'corrupting' sea of Horden and Purcell (Abulafia, 2002: 20). They have summarised their approach as a four item model – 'of risk regime, logic of production, topographical fragmentation, and internal connectivity' (Horden and Purcell, 2006: 722–740). Within this quartet they ask us to appreciate better the role of marginal lands (including the sea) in helping to offset the potential fragility of the famous Mediterranean trio of crops – wheat, olives and vines. The uplands, the wetlands and the inshore waters gave flexibility to the agricultural economy and the food supply, helping to reduce risk. Similarly, the forests were a rich source of a variety of resources, not just fuel and timber. In Mallorca, for example, at a local scale the *hortes*, gardens and allotments around the island's towns, especially those founded by the 1300 *ordinacions*, could be seen as an important buffer in the food supply system providing vegetables and fruit to a largely non-meat eating society. When these supplies proved inadequate, longer distance trade – especially in cereals – more than supplemented local shortfalls.

David Abulafia takes a very long view of the Mediterranean from prehistoric to modern times, but it is a view that is 'resolutely (of) the surface of the sea itself, its shores and its islands, particularly the port cities that provided the main departure and arrival points for those crossing it' (Abulafia, 2011: xvii). Faruk Tabak's Mediterranean concentrates on the 300 years from the late sixteenth century. The value for the landscape historian of his perspective is the focus on the economic shift from the plains to the hills, the move from grain production and trade to the trio of tree crops (almonds, carobs and olives) and vines (Tabak, 2008). As context for the island of Mallorca, both have a lot to offer. This story of Mallorca's landscape will take this more microregional view, namely that while the Mediterranean provides a context, its landscape is made up of too many exceptions and differences for any general rules to enable us to file it all too easily and lazily as 'mediterranean' in the Braudelian sense, as we shall try to set out briefly at the end of this chapter. Mallorca is more *in* the Mediterranean then *of* it.

Before then, however, there is one other contextual aspect to which we must give brief consideration that is important to the landscape historian. This is the idea of, for want of a better term, 'Paradise Lost'. There is a longstanding view that mankind had taken a rapacious view of the environment in the Mediterranean lands, exploiting its natural resources for economic gain irrespective of the ecological consequences. According to Grove and Rackham this has four origins: Renaissance artistic views; floods – seen by many as abnormal features rather than as extreme but normal events which, it was said, led to loss of forest cover and increased soil erosion; the role of trees – thought to increase water vapour and hence rainfall, therefore tree loss was believed to lead to a reduction in precipitation; and lastly, experience from newly discovered oceanic islands that were known to lose much of their vegetation once settled, a process that was thought must also have happened in the post-Classical world. From then on the supposed, but never proved, forest cover of tall trees was, it was said, stripped away by logging for shipbuilding, domestic construction and charcoal-making or consumed by herds of goats. The Mediterranean landscape was seen as an example of large-scale degradation with the present scrubland and scattered trees as a debased form of forest. Where 'learned visitors seeing scattered trees, are irresistibly tempted to interpret them as the remains of a forest: in reality they may be trees that have sprung out of *maquis*, heath or cultivated land, or maybe the latest of several generations of savannah trees' (Grove and Rackham, 2001: 214). McNeill believed that as late as the nineteenth century, 'The value of timber generally exceeded the annual income of forests by three to five times in Spain, so only the patient and unindebted refrained from logging their new lands', as though somehow this attitude was and had been universal (McNeill, 1992: 262). Grove and Rackham are at pains to point out that the so-called natural vegetation was unlikely to have consisted everywhere of 'forest' but that tree savannah, especially in lowland areas, was more likely. They also remind us that vegetational change, including tree cover, is also the product of climatic variations over the last 12,000 years; a 'degraded' landscape, if degraded it be, is more likely to be the product of natural forces.

Landscape, whatever its origins as a word, is to do with 'seeing' the material, real world translated into an image via our perceptual mechanisms, themselves fashioned by culture and experience. To what extent, then, is the small island of

Mallorca known to millions of tourists as a *particular* place, typical of the wider Mediterranean? How far is Mallorca the epitome of things Mediterranean? Or is it so much like, say, a Greek island, the Adriatic coast of Italy or Croatia or the Costa Brava that to all intents and purposes we see it in the same light as other such places? How far can we generalise about the Mediterranean from the Mallorcan experience? Are they in any way synonymous terms?

Given the present author's background it might be expected that geography might predominate in any such discussion, and to an extent that is true, but when dealing with landscape it becomes more and more necessary to see 'mediterranean' as a cultural construct, a trait we have already pointed to in the earlier discussion on landscape. To most Europeans the Mediterranean is closely associated with Classical antiquity, particularly that of Greece and Rome. The Mediterranean was the setting for the exploits of the Gods, the creation of sea and land empires of the Greeks and Romans, the context for philosophies from Socrates to Cicero. Reading the Classics, especially in the eighteenth and nineteenth century English public schools, meant also studying the geography of the Greek and Italian peninsulas, and of Anatolia (Arnold, 1849). Indeed, it is possible to trace the development of geography as a teaching subject partly to this origin. The geography of the Mediterranean was also the essential context of the Grand Tour of the English gentleman. Against this, the study of Arabic cultures was much less associated with the Great Sea, perhaps because the Arabs who conquered Spain were much less sea-goers than their predecessors, their Berber armies from North Africa even less so. Today, the Mediterranean is synonymous with the culture of tourism dating back to the area as a winter destination for north Europeans in the late Victorian and Edwardian eras. Now, of course, it means mass tourism. The cultural spectrum ranges from *mare nostrum* to Club Med.

Physically we take the Mediterranean to be a basin, an enclosure with one outlet to the west, but to most of us it is a seaway, a throughway, a surface feature upon which people, goods and ideas are moved. For many this movement was, until recently, perceived by many European historians and geographers as an east and west movement; it was an imperial routeway especially to India and South East Asia, but for the French and Spanish, and later for Italians, it was also a north–south movement to Algeria, Morocco and Libya. We shall show that in the case of Mallorca, north–south movements were especially important, particularly for trade. In fact, the Mediterranean is a series of basins, some graced with the names of seas such as the Tyrrhenian Sea, the Alboran Sea, the Adriatic, etc., others simply inlets, even if on a large scale, such as the Golfe du Lion. Essentially there are two basins separated by the Straits of Sicily, and it is with the westernmost of these that we shall be most concerned; but it would be wrong to assume that connections between the two were minimal. The area of the western basin is 821,000 km^2 – about half the size of the eastern one. There are 32 islands in the former as against 130 in the eastern basin. The average size of island in the west is much greater at $2130\,km^2$ vs. $297\,km^2$ in the east (Barceló Pons, 1972: 8). As an enclosed sea elongated east–west, the Mediterranean has two rims to north and south with the islands – and the cliché is fortunately unavoidable – as stepping stones between them, Mallorca being a major one in the western basin. While obviously true that 'distances were narrowed by the presence of islands' (Abulafia, 2003: 26), what became more important in cultural terms was that these stepping stones also became places where cultures came into intimate contact with each other. Nearly all the islands – Cyprus, Crete, Sicily, Mallorca – exhibit traits derived from many cultures, many of which persist in the landscape today. The idea of *cultural overlay* is as important to the landscape historian as the notion of *palimpsest* outlined earlier. Such cultural intermixing usually took place in sea port locations, especially in the larger cities such as Palma. It also meant that from an early date Mallorca established consulates on the North African coast to facilitate this trade, further enhancing the possibilities of cultural mix (Abulafia, 2003: 21). This is not to advocate the 'melting pot' principle necessarily, although often in Mallorca and elsewhere a dominant political force often over-rode a less powerful one, subjugating the latter to the tenets of the former. For example, the *mozarabs* were a product of Muslim domination over Christians, the *mudejars* a product of opposite forces. The *xuetes* of Mallorca are remnants of a once powerful but then subjugated Jewish culture. Thus, these subsets or amalgams may also have had their own landscape effects. We have to remind ourselves again that chronologically speaking a new culture did not completely replace an old one, however much ethnic cleansing went on. This would have had a spatial expression, creating not only a historically 'lumpy' landscape but a geographically diverse one too.

The key to this is to recognise the respective and related roles of time and place in relation to human activities: *flux* is a useful descriptor, a somewhat distant cry from Braudel's continuity. Rather than seeing the landscape of Mallorca as 'degraded' or 'ruined', a view based on society's disrespect for the environment – or in David Abulafia's words, the absence of a 'dark history of mankind corrupting a terrestrial paradise by environmental incompetence' (Abulafia, 2003: 21) – we must learn to see it as a constantly changing one in which there are no 'eternal verities' of value, but rather ongoing appraisals and reappraisals of environmental constraints, including, where necessary, recognising that environments are not fixed or immutable; they are quite rapidly changed by technology and human ingenuity. Irrigation from artesian sources, for example, transformed Mallorcan agriculture in some areas, giving us the windmill as a prominent landscape feature. Many of the beaches of Mallorca beloved by hotel builders and tourists alike have come and gone with the seasons and with changing sea levels and storm patterns. What the tourists see and sunbathe on is often the product of the JCB and the bulldozer hard at work just before they arrive (Buswell, 2011: 17–25). The extensive almond blossom displays of February really only date from the eighteenth century, the recent revival of vines in the lowland landscape around Binissalem and Felanitx date from only the last three decades or so, *phylloxera* having destroyed many of the originals in the 1890s. The potato fields around Sa Pobla are a product of a twentieth century cooperative movement; for hundreds of years before much of that area was ill-drained land or under cereals.

If we can give less credence to 'ruin' and 'degradation' for preceding centuries, is it possible to ask whether the sudden and massive impact from mass tourism from the 1960s onwards has, in fact, now led to that condition – degradation, but degradation of a different and complex kind? Sustainability is a modish term, but if we take Braudel's view of the *longue durée*, does it mean we should expect the landscape of tourism to continue, decline or be modified? It has become such a powerful political, economic and social activity that clearly, in the short run, all efforts are being directed to its sustenance – but then, that was what was probably said about the new-fangled sheep economy of the late medieval era or of Sóller's orange groves in the 1850s. Landscapes are rarely laid waste deliberately, despite greed and insensitivity, especially on islands. After all, there's nowhere else to go. The value of landscapes reflects the dynamics of shifting natural and human forces within a Mediterranean locational setting so that its history is perhaps akin to the economists' search for that chimera, the 'long run'.

Within this Mediterranean setting, how far is Mallorca typically Mediterranean? Are there certain characteristics of landscape in the island that are distinctive, even the degree to which some of them might be considered unique? It might be convenient to simply list those characteristics of Mallorca's history and geography, which at a superficial level at least, appear to contribute to a Mediterranean landscape. Mallorca is:

- An island, and suffers to a greater or lesser degree from isolation and separateness;
- its geology and climate appear to fit the 'classic' patterns of the Mediterranean basin;
- it developed a Bronze Age culture;
- it was conquered and settled by the Romans and absorbed into their empire;
- it was subject to Muslim invasion;
- it is an example of Christianisation following the Aragon/Catalan conquest;
- saw the imposition of feudalism, especially its landholding structures;
- experienced the reform of land holdings and redistribution of land from the sixteenth century onwards;
- witnessed the development of a new tree-based agriculture in seventeenth and eighteenth centuries;
- saw late industrialisation;
- late nineteenth and twentieth century emigration to ease pressure on local resources;
- exhibited the primacy of its major urban settlement for many centuries;
- developed early tourism; and

- mass tourism from the mid-twentieth century;
- has subsequently developed a post-touristic economy and culture; and
- experienced population change via emigration and immigration.

To those who have studied other parts of the Mediterranean or the area as a whole, this list will have a familiar ring. However, when examined closely nearly all the detail in these 'events' appears much more 'mallorcan' rather than 'mediterranean'. Again, and without going through the whole of the previous list, another interpretation might begin to correct this view:

- Mallorca is very much of the *western* Mediterranean in its geology and climate, quite different from the eastern basin.
- Its Bronze Age culture was late in settling and much of its built form (the *navetes* and *talayots*) quite different from other relict features from elsewhere in the Mediterranean, even different from neighbouring Menorca.
- The Roman settlement had little impact on the cultural environment, especially in the countryside.
- Muslim settlement came late in the early tenth century and lasted for a much shorter period until 1229. It was largely Berber with origins in the Almoravid and Almohad societies rather than true Arabic;
- although at the frontier between Christian and Muslim cultures there is the paradox of trade by Mallorca with both;
- feudal land-holding patterns and cultures particular to Mallorca, especially the early introduction of share cropping and 'cash and kind' rental. Its patterns of inheritance are quite different from many other Mediterranean areas including nearby Catalunya.
- The conversion to sheep ranching latifundia was not on anything like the scale of the Peninsula, and not solely because of environmental constraints.
- *Possessions* (landed estates) emerged as new engines of agricultural reform from the sixteenth to nineteenth centuries.
- It was not on the Grand Tour itinerary, therefore north European perceptions of Mallorca were barely influenced by the usual search for 'classical' landscapes and artefacts.
- Industrialisation was not based on north European models that included waterpower, steam, coal, iron and steel, textiles and railways, but on small-scale factory development and home working together with new products such as shoes.
- The domination by the primate city (Palma) was much greater in demographic, social, political and economic terms. For long periods 'Palma was Mallorca'; there was only poor development of the lower parts of the urban system.
- Mass tourism has had a more rapid and greater impact in scale or proportionate terms, and came earlier than almost anywhere else in the Mediterranean, leading to a narrow economy based on the tertiary sector, but nonetheless making Mallorca one of the richest places in the Mediterranean.

- Immigration from the 1960s to the present day has transformed the island's demography.

Mallorca may not be unique, but is sufficiently different in its geography and history to qualify it more as one of Horden and Purcell's *microregions* rather than part of Braudel's overarching unity of the Mediterranean environment and historical experience. It is a theme we shall need to explore in landscape terms.

3

The physical basis of the landscape

Del llim d'aquesta terra sa vida no sustena;
Revincla per les roques sa ponderosa vel,
Te pluges i rosades i vents i llum ardenta,
I, comun vell profeta, rep vida i s'alimenta
De les amors del cel.

The earth of this land gives it no sustenance;
Twisting into rocks, feeding on
showers and dew and wind and scorching light;
And, like an old prophet, winning food and its life from heaven's breath.

(From *El Pi de Formentor*, Miguel Costa i Llobera, 1854–1922, author's free translation)

Introduction

For a landscape history such as this, the need for a detailed account of Mallorca's physical geography is so much less than a modern geography might require. In such an eclectic field the need for such a section has varied largely according to the background of the writer. Curiously, our 'hero' in landscape history, referred to earlier, namely W. G. Hoskin, did not include a physical introduction to his pioneering work on England (1955) but then, he was a historian. For this author its inclusion is important because the geology, geomorphology, weather, climate and vegetation, and the ways in which these combine in an ecological sense, provide more than a mere backcloth to the passage of human events. If 'landscape' is about seeing – a term derived from art – then, depending on distance and perspective, the most obvious element of any scene is its physical appearance. The shape of the rocks, the relief, the variations in the colour of the soil and the vegetation all strike the mind's eye first before the covering and moulding made by human endeavour are taken into account. Initially our perceptual mechanisms see the hills, the mountains, the coasts and the woodlands as foreground; the socially constructed elements come into view as we focus in on the details. So, from two points of view the physical environment is important to the landscape historian or historical geographer: it poses constraints and offers opportunities to society over a long time period and it forms an important part of our perceptions of the landscape today.

One hundred years ago no-one writing a geography of somewhere like Mallorca would have thought of omitting a chapter on the 'physical basis' of the island. By the mid-twentieth century geographers had learned the lessons of their French predecessors and admitted to their view that the physical environment offered opportunities to human action rather than determining it: determinists had given way to possibilists. When Jean Bruhnes wrote about the geography of Mallorca in 1947 (Brunhes, 1947: 545–560) and when a later compatriot, Jean Bisson (Bisson, 1977: 287–292), submitted a doctoral thesis on the island in 1974, both nonetheless included a section on the various elements of the physical appearance. Geology, geomorphology, drainage, climate and weather, soils and vegetation all figured as an essential part of the description of Mallorca, if not always as part of its analysis. Earlier, the British geographer E. W. Gilbert was perhaps the first such to write a paper on its human geography in English (Gilbert, 1934),[1] although amongst an earlier generation Clements Markham (a former president of the Royal Geographical Society (1893–1905) had, surprisingly, included only a passing reference to this aspect in his book on the story of Mallorca and Menorca as early as 1908 (Markham, 1908).[2] Nearly all of Mallorca's own practitioners have followed these leads, despite the lessening credence given to the role of the environment in shaping human action on this Mediterranean island. Rosselló's masterly comprehensive survey and explanation on the development of the south and south-east of the island and Flórez's account of the Tramuntana followed their lead, as did Pere Salvà's massive doctoral work on the same area (Rosselló Verger, 1964; Ferrer Flórez, 1974). In much more recent times Rullan Salamanca's attempt to explain Mallorca's regional divisions from long historical and later political perspectives does much the same thing (Rullan Salamanca, 2002). It is not unreasonable to ask, then, why have many more recent writers felt it necessary to preface their spatial descriptions and analyses with sections on the physical basis of Mallorca when in terms of the then evolving or current paradigms, the idea of environment shaping human action had long since passed?

The Mediterranean context

Perhaps the answer to the question of the physical environment's significance lies in the fact that Mallorca is a Mediterranean island. It is in the nature of that setting that we must examine the role of the environment in helping to explain the characteristics of Mallorca's landscape history. A simple but largely false beginning might lie with the notion of the immutability of the landscape in the longue durée that we have touched on in the previous chapter, the supposed timelessness of the Mediterranean scene. Rather we should see the setting as one most certainly changing over time, not constantly and slowly, but often dramatically and over very short periods. Examples range from the volcanic activity of the Mediterranean lands occurring over millennia of geological time to the destructive action of the frequent flash floods happening over a matter of hours. Rather than evolving slowly along a curve, the Mallorcan physical environment is much more the product of short, sharp, step-like changes.

Three aspects of the so-called 'universality' of the Mediterranean environment might be listed: the largely limestone and sandstone geology accompanied by the Alpine folding of the Tertiary era giving rise to a variety of relief types, including mountains and plains, but upon which constantly running water today plays only a small part; the distinctive climate and weather pattern of the area with summer droughts and often severe autumn and spring storms and lastly, a 'natural' vegetation of pines, holm-oaks and garriga or maquis that has been much modified by human action over at least six millennia of occupation. In very general terms, these three have become interrelated to produce the characteristic landscape of the Mediterranean with its hard limestone mountains whose lower slopes are clad in oaks and pines, its red soils in a series of plains and basins and its dry valleys subject to irregular floods. Added to them, in the case of most Mediterranean islands, might be the post-glacial fluctuations in sea level which have given rise to particular coastal features of coves and headlands that in the late twentieth century took on a new resonance with the development of tourism. However, these are simple generalisations, and as discussed in the previous chapter, it is the subtle, place-to-place variation to this general picture that holds the key to explaining the geographically specific responses of society to the physical and natural environment. In terms of so many aspects such as the degree to which relief is accidented, the considerable spatial variation in rainfall totals and frequencies, soils and water supply for farming, coastal features for harbours and the related urban development, the density and tractability of lowland scrub and the disposition of woodlands for timber and fuel – in all of these the environment is relevant. The challenge for the writer is the breadth and depth to which he or she wishes to go in order to assist in narrating the story of Mallorca's landscape history. For this work a fairly simple solution has been adopted; those who want more detail are referred to the excellent first part of Onofre Rullan's *La construcció territorial de Mallorca* (Rullan Salamanca, 2002: 27–105) and the *Atles de les Illes Balears* edited by Pere Salvà.

Rocks and relief
Any introduction to the physical geography of Mallorca (Fig. 3. 1) begins with the simple three-fold division of the island into the uplands of the Tramuntana in the north and west and the Serres de Llevant in the east with a central lowland area between. The basic geology that in part gives rise to this division is dominated by rocks of the Mesozoic era, that is from about 280 million years ago to 100 million years ago. The core of the higher lands is made up of Liassic rock of the lower Jurassic and the more recent series of the Cretaceous. These are sedimentary rocks principally of limestones, some massive, some much softer, and sandstones. They make up the highest peaks such as Puig Major (1445 m) and Massenella (1367 m). The core of the intervening plain is essentially of the Miocene period, and is made up of chalks, marls, clays and even the remnants of some tertiary volcanics. This simple picture is further complicated by two other features: a dense cover of

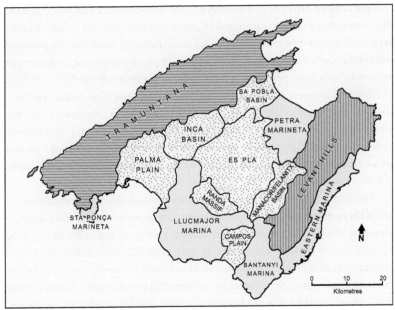

Figure 3.1 Map of physical regions. Source: *Atles de les illes Balears*, 1995.

quaternary deposits on the eastern flanks of the Tramuntana and the inland sides of the Serres de Llevant, and two large plains or marinas to the north and south of the central plain (Es Pla), with a third one from the southern tip of the island and along the east coast composed of the marinas of Santanyí and Llevant. Here the under-lying rocks are essentially upper Miocene overlain with Pliocene and Quaternary deposits. This oversimplified picture of the geology of Mallorca is sufficient for our historical purposes. It forms the basic structure upon which the detailed landforms that have affected the human responses have developed. Geomorphology and soils have been more responsible than solid geology for fashioning the ways in which society has used the land.

Perhaps the most noteworthy surface characteristic is the karst landscape pro-duced from the calcareous lithology (Gines and Gines, 1989). One of the first fea-tures that strikes the visitor is the absence of running water over most of the island, a fact not solely attributable to the dryness of today's climate, but to the porous and soluble nature of the island's rocks. Water has nonetheless perhaps been the most significant shaper of Mallorca's surface, though not in the usual way of valleys cut by running streams and rivers. Surface water has soon percolated down through the limestones and chalks and then shaped enormous underground features such as caves, large caverns and underground valleys. Many of these have subsequently collapsed to open up the surface and produce a sharp and broken relief interspersed with broad depressions. This superficial karst landscape (*lapiaz*) is typical in the mountain area between Pollença and Soller, where there are longitudinal valleys running roughly NE/SW and a series of canyon-like valleys orientated at right angles on either side of the principal ridges. Similar features can be found in the Serres de Llevant, but on a reduced and more rounded scale. On the Miocene areas of the south and east coasts another form of more gentle karst landscape can be found, made from post-Alpine flat tablelands partly made up of former reefs, which are cut across by shallow dry valleys running east and south to the sea. These are continuations of much steeper-sided valleys or *barrancs* debouching from the Serra. This area contains some of the most dramatic underground cave features in the island such as those of Drach and Hams (near Porto Cristo) and Artà that are well known to the tourist, but there are huge areas of similar structures known only to the speleologist.

The only area where karstic influences on the surface are not pronounced is on Es Pla proper, where more recent Quaternary deposits mask or bury them. In some ways this central area is the most complex geological area in Mallorca, giving rise to a variety of landforms including quite high hills such as Randa (544m) and many smaller hills and undulations. However, the varied geology has helped create an equal variety of soil types and colours that together add a distinct veneer to the visual appearance of the landscape. This is one of the areas that developed the large estates (*possessions*) formed form the eventual break-up of the original medieval landholdings of those that had assisted Jaume I in his conquest of the island in 1229. Physically the Pla is much more accidented in relief than its name suggests. Many Mallorcans see this central area of their island as Mallorca profunda, the 'real' Mallorca, as distinct from the alien urbanisations around the coast, an area which some say should be preserved rather than conserved and integrated into contemporary Mallorcan life (Picornell and Picornell, 2008: 205). In the fifty years since 1960 it has been that part of the island perhaps most abandoned by agriculture. As a result the vegetational patterns now include large areas of naturally generated *pinus halepensis*, Aleppo Pine.

The coastline of Mallorca, naturally, requires some comment from the physical point of view, since it is the locus of the modern tourist industry and is its principal natural resource. To the north and west it consists primarily of steep cliffs descending almost vertically into the sea in the central section; only the harbour of Sóller gives any real access inland. There are small coves where torrents pour out giving rise to small beaches such as at the mouth of the Torrent de Parais, and on a more gentle scale at Cala Sant Vicenç. All other coasts exhibit various aspects of less dramatic coastal scenery. To the south the coast of the marinas has low-rise cliffs of former coral reefs fronted by long beaches such as Es Trenc. To the north the two great curving bays of Pollença and Alcúdia have narrow sandy beaches that are the product of post-Flandrian changes in sea level, which in fact affected all of the island's coastline.

One final landform that must be noted is the albufera, a lagoon-like feature that is cut off from the sea by dune formations. Such wetlands are curiously characteristic of the Mediterranean coastlands. In Mallorca's case the Albufera is located in the north of the island in the Bay of Alcúdia with a smaller version – the Albufereta – in its neighbour, Pollença Bay. They are the result of tectonic and sedimentary processes taking place throughout the Alpine era during the lower and upper Miocene (Fig. 3. 2). The area between the two ranges of Tramuntana and Llevant subsided, and during the upper Miocene and Pliocene began to infill with carboniferous limestone material eroded from the mountains. The waters that accumulated were then enclosed by sandbars related to the eustatic sea level variations associated with the Quaternary glaciations, particularly during the long Pleistocene era. During the Lower Pleistocene the whole area was flooded by a huge delta that was located far to the east of today's coastline. In the Upper Pleistocene a series of sedimentary deposits and beaches was laid down. Quaternary oscillations throughout the Riss, Würm and Flandrian eras gave rise to aeolian (wind-borne) deposits formed from fragments of molluscs and coralline algae (Rhodophyceae) forming coral reefs. During the Tyrrhenian and Flandrian transgressions sand dunes were laid down, initially on the edges of the basin with marshland developing behind. During the Flandrian the arc of sand dunes was completed, but further inland than the present line. More extensive and continuous marshland developed behind this barrier from about 50,000 BP. Offshore bars have gradually moved landwards as further terrestrial deposits washed down from Serras have filled in the marshlands from the west, which have gradually

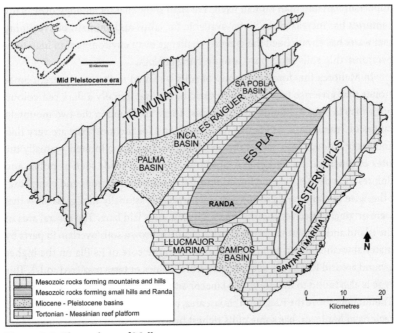

Figure 3.2 The geology of Mallorca.

been colonised by reeds (phragmites) (Fornos, 1995). In the nineteenth century 30–40% of the Albufera's surface was open water; today it has been reduced to about 3%. Chapter 9 will explore how this physical landscape has been changed by many centuries of human action.

Soils

Soils are a product of the local base rocks and of superficial deposits laid down by various depositional forces, followed by in situ weathering and reworking by surface water and, in some cases, wind. In a land dominated by sedimentary geology, one of the most striking features of the island's surface is the large variety of colours rendered by the underlying rocks. In any transect across the farmed areas of central Mallorca after the autumn or spring ploughing, one is immediately struck by the variety and rapid transition in soil colour from the bright white chalks, the brown and ochre marls, the red soils of the limestones and the grey clays, fading to the pale yellows derived from sandstones. At certain seasons, soil forms a significant element in the landscape's palette.

When examining the contribution of land-use to the landscape the important point to make about the role of Mallorca's soils is their poverty; in this it shares many of the characteristics of many Mediterranean countries. It is a combination of soil types, depth and distribution that together contribute to their poverty rather than their dryness and proneness to erosion; a lack of humus is often a characteristic. For agricultural purposes water supply and availability is a more important variable than rainfall itself. We shall show that technology over many centuries has increased the water available for crops and animals, but retaining that water has always been the greatest challenge in an environment of high temperatures, thin soils and a porous underlying geology.

In Mallorca the dominant group of soils is related to the omnipresent limestones that give rise to the characteristic terra rosa, literally a dark red colour, but which is poor in humus and nutrients. These are found in the two mountain ranges of Tramuntana and Serres de Llevant, and for the most part are very thin and easily eroded. Only where the intramontane basins are located – usually the sites of downward drainage and sinkholes of various kinds such as poljes and dolines – are deeper, richer deposits found that can support crops and ploughing; otherwise the land tends to support only trees, occasionally thin grasslands that were originally used for transhumance grazing, or is laid bare. The central area of the island and the eastern marinas exhibit mostly brown soils overlain in parts by material washed down from the limestones. In the core of Es Pla on the higher ground around Randa are to be found xerorendzines or terra rosa (red soils). This type is also found to the north of Manacor and on the upper slopes of the central Tramuntana above the Raiguer. This last area, besides being one of the best-watered regions of Mallorca, has some of its richest brown soils derived from Quaternary and recent erosional deposition. Such soils are also found in the Sóller basin. The post-Flandrian transgressions of the northern and southern bays have yielded quite different soil types based on their wetland origins, so that they are amongst the few areas that have considerable humus in their make-up.

It's weather not climate

There is a strong case for the argument that says that Mallorca's weather, rather than its Mediterranean climate, is responsible for the effects of temperature and precipitation on its landscape. While the broad features of hot and dry summers and the relatively warm and wet autumn and spring periods may lay down a framework, the detailed effects are derived from the more extreme events in particular years or seasons. Drought, for example, is a frequent occurrence, but by no means a regular one. Average rainfall figures mean little if there is great variance in its annual distribution. The autumn and winter winds from the north across the Tramuntana are fairly regular and predictable, but it is the sudden and often vicious cyclone coming in from the south and south-east that uproots trees and tears off roofs. There are, in addition, longer fluctuations in the climate patterns, such as the Little Ice Age of the fourteenth to eighteenth centuries and the phenomenon of global warming in our own times. These have had – and are again beginning to have – a marked effect on agriculture, woodlands and water levels – all influencing how the land was fashioned and used. It would be possible, of course, to chart the changes in Mallorca's climate over the last 6000 years and note its impact on landscape features. The trick would be isolating those features in the contemporary landscape affected by climates in times past from the effects of more recent climatic shifts and meteorological patterns, but time and space prevents that here.

Precipitation since the middle of the Little Ice Age has been declining until fairly recently as the Mediterranean climate has got drier. This has meant that many crops, especially cereals, have found it increasingly difficult to flourish, and as we shall see, this has had a major effect on the island's ability to feed itself. It has encouraged, but by no means caused, a movement away from field crops towards tree crops, particularly deep-rooted ones, and to increase the area under irrigation growing higher value crops such as animal fodder, fruit and vegetables that can, in part, recoup the cost of artificially adding water to the land. This process is not entirely new; its economics underlay much of the medieval Arabic agricultural economy of Mallorca. Over the centuries, however, it is probable that water supply and its retention played a much less important role than the poor quality of island soils in determining the long-term productivity of many classes of land use. As far as rainfall is concerned, what mattered was the annual variability in the amounts of rain; drought, or more accurately the failure of the rains in spring and autumn, is only a major problem if it persists over many years. In terms of the 'natural' vegetation, such a long-term drying of the climate was probably responsible for the decline of tree cover over large areas of the island, permitting the spread of garriga, especially in the lowlands.

What is really important in present-day rainfall patterns is the deluge (*aiguade intense*), the sudden and rapid decanting of huge amounts of water onto the surface, often in a matter of hours. In a study of twentieth century heavy rainfall Grimalt, Laita and Ruiz examined the severity of such events in the east of Mallorca (Grimalt Gelabert *et al.,* 2001: 29–39). On 3 and 4 October 1957, for example, 400 mm of rain fell in 24 hours in Santanyí and 276 mm fell in the same period at Son Crespi Vell on 2–4 November 1943. In this eastern area about 60% of such intense events occurred in the autumn (especially in the first half of October) and only 16–17% in the spring. Whereas the Tramuntana is generally the area of highest rainfall, these data show that extreme events can occur outside the principal mountain range, the resulting floods in Ses Salines in 1957 being a good example. In nearly all such cases they lead to surface damage and normally dry valleys are inundated and the man-made torrents can rarely cope with the volumes of water involved. On the east coast where the barrancs meet the plain, enormous quantities of eroded debris are suddenly deposited, mostly made up of stones and boulders rather than soils (Fig. 3. 3). Of course, it also has to be remembered that with so much of Mallorca being made up of heavily fissured limestones, a considerable amount even of excessive precipitation is soon lost from the surface.

Figure 3.3 Seasonal flooding in Eastern Manacor.

The side effect of such deluges is erosion. It is a complex problem in which tectonic and lithological factors are thought to be more important than soil type and vegetation cover. Two historical examples may be cited of the more dramatic kind of erosion in the form of landslides. In 1721 at Biniarroi near Mancor de la Vall a cultivated area of more than 300,000 m² slipped, moving in excess of 7 million cubic metres of rock, soil and debris. In more recent times another example could be found at Fornalutx, where in 1924 a landslide destroyed 150,000 m² of olive-growing land (Mateos Ruiz, 2010) Although an age-old problem, it is probably true to say that it has worsened in some areas in the last fifty years; a common victim in the last few decades has been the expanded road system. Precipitation events have become more intensive, if less frequent. At the same time, some land has been intensively farmed and has become more mechanised, resulting in larger, more open fields, much of this the result of subsidies, also leading to greater soil erosion. In the less intensively farmed areas and those areas abandoned by farming, erosion has probably reduced. The real changes in the superficial landscape occur in only three or four extreme rainfall and erosional events per century (Grove and Rackham, 2001: 268). Attempts to manage some of them by, for example, investing large sums in constructing flood prevention schemes as in the *torrents* of Mallorca are now viewed by many as often making matters worse. Earlier generations who did not have the 'benefit' of modern flood prevention technologies devised other management strategies that worked *with* the environment.

Drought – the other high risk extreme event – is a common occurrence in Mallorca. If persistent it often leads to economic crisis as crops fail and the food supply dries up as, for example, in the mid-nineteenth century between 1845 and 1850, when the bishops led large processions calling on God to intervene and send rain. (He didn't, of course, but when it did eventually rain the bishops naturally claimed the credit!) Droughts, though intense, are often short-lived, as in 1945 when one lasted from February to July; but it followed a year of excessive rainfall. And because these kinds of events are part and parcel of the meteorological scene in all Mediterranean countries, society has learned to cope with them; they are not seen as exceptional but as frequent, if not regular, happenings.

Few of us associate the Mediterranean with snow, but falls occur frequently, largely on higher ground; the Tramuntana in Mallorca is no exception. It does, however, take the Mallorcans by surprise each time. Although snow itself is rarely damaging, the low temperatures that accompany it often are. Frosts are particularly damaging to the tree crops of almonds and various fruits. The accumulation of snow and ice on high ground was long exploited in the days before mechanical refrigeration; the construction of snow houses (*botigues de nue*) in the mountains to supply Palma with ice can be traced back to the fourteenth century. They became as much a feature of the landscape of the mountains as the *sitges* (charcoal making) and the *calciner* (lime burning) sites.

Water on the surface: water down below – the ongoing challenge of water supply

The section above on climate mentioned the problem of water supply in Mallorca. In the light of the relative unpredictability of rainfall it is not surprising to find that throughout the island's history obtaining and retaining water has been amongst the greatest challenges to its technological ingenuity. As this history of the landscape unwinds we shall have recourse to mention this fact on many occasions. Of all the elements acting as a 'constant' in the landscape, water is perhaps the most persistent, no more so than today when an island population that might reach 3. 0 m at the peak of the tourist season has to be fed, watered, bathed and sewered. It would surely be possible to write a cultural history of Mallorca devoted solely to this topic.

In the early history most energy went into obtaining a year-round supply of water, distributing it to farms, towns and villages and storing it. The Romans may have begun such systems, but it was Arab technology originally based on Roman engineering practices that contributed most, continued by the Christian invaders and settlers right through to the modern age of the nineteenth century. The *font*, the *qanat*, the *aljub*, the *bassa*, the *safareig*, the *sèquia*, and the *noria* have not only contributed much to the landscape but to the island's vocabulary too (Fig. 3. 4). From the mid-1800s to the present day new features (and words) were introduced: the steel windmill, the diesel and electric pump, the reservoir, the pipeline and even the water-tanker ship. Of whatever age, these usually small-scale, man-made objects associated with the management of water supply figure prominently in the landscape of Mallorca, and even more in its economy and social organisation.

Water is drawn from two sources: the surface, from springs, streams and rivers and natural lakes and ponds, and from underground, from aquifers or subterranean lakes. Given the geology we have described, superficial sources have been sporadic in both time and space, their very presence dependent upon seasonal variations in precipitation and the level of the water table. The principal challenge from such sources has been to capture the water when it appears or devise methods such as the Arabs' *qanat* to make its appearance last longer (see Chapter 6). In Mallorca, rivers and streams only run intermittently. Seasonally their presence is determined by rainfall patterns and especially by the deluges mentioned above. Even in the wetter seasons, rain is by no means the regular and frequent occurrence it is in more temperate climates. Streams and rivers are also sporadic spatially on Mallorca's largely permeable surface, often disappearing down sink holes or into underground caverns, sometimes reappearing downstream, sometimes not. Inspection of any map of Mallorca's drainage system will show the valleys and river beds, but only rarely is water to be found in them all year round. In the upland areas such surface waters are, naturally, a more ever-present set of landscape features: rainfall is very high here, the water tables are near the surface, and the sound of springs and running water is common, but the broken relief and the poor quality

Figure 3.4 A noria, an animal driven means of raising water, reconstructed at the Museum of Mallorca, Muro.

of the soils means that the water often cannot be used here, so that the mountains become the gathering grounds for water to be used elsewhere. Historically this was achieved with the building of long-distance sèquias such as those in the Coanegra valley or in the upper reaches of the Massenella catchment. In more recent years it has meant the construction of dams and reservoirs such as Cuber and Gorge Blau, linked by pipeline to the major centres of consumption.

Subterranean supplies of water are much more complex, thanks to the varied geology, which can only be understood from detailed analysis of drilled cores, and to the geography of the underground drainage system. Although there are some natural underground lakes which act as reservoirs, most water occurs largely in the interstices of permeable rocks that go to make up aquifers. From these, water may be forced to the surface by natural pressure or, more likely today, has to be either manually raised or pumped to the surface. Such aquifers occur widely over the lower lands of Mallorca, especially in Es Pla and the north and south *marinas*. They have been variously accessible by a changing series of technologies developed over time, but all dependent upon some kind of 'well', or shaft, sunk into the aquifer.

The well is a most significant cultural feature in the Mallorcan landscape and complex rules have governed its initial sinking and the subsequent use of its water.

They are to be found widely scattered over rural and urban areas, many now largely abandoned as municipal piped water supplies much of the island for domestic consumption. In rural areas where water was required for irrigation purposes, the mule- or donkey-driven *noria* was most common. In towns, villages and the isolated farm the simple well with a bucket on a rope was the order of the day. In both cases raising the water was only part of its management; distributing it to the fields or to the households of a settlement was governed by much more complex rules set down and enforced by a local democracy, details of which unfortunately cannot be entered into here.

But the well taps into a common good, and like all such commons it suffers from the usual 'tragedy' which states that the resource in question will not be equitably used as demand rises and the regulation system that stood it in good stead for centuries becomes less able to manage and, in this case, water becomes overexploited (Hardin, 1968). This has become especially true in more recent years as demand has grown for agricultural irrigation (by far the largest consumer of water in Mallorca, much more so than even the island's rapacious twenty golf courses or its nine million tourists!) and domestic and industrial consumption. At the same time better technologies have been introduced to capture more and more. Even individual households in the countryside now believe that the water in the aquifers beneath their land is a common good, and somehow unconnected with the supply to wells elsewhere. One of the most common sounds in the Mallorcan countryside in recent years has been relentless dawn-to-dusk drilling as the better-off population disperses away from the cities and towns and demands its swimming pool alongside its washing machine and its three bathrooms; rarely are they prepared to pay the real cost of the water supply. A similar selfishness is practised by farmers and golf course promoters and is far removed from the regulated water supplies of previous generations.

A second problem is the fact that excessive depletion of the aquifers' supplies over the last 50 years in some areas has lowered the water table to such an extent that much more powerful additional pumping becomes required. Amongst the most potent icons in the Mallorcan landscape are the steel windmills of the nineteenth and early twentieth centuries. These could raise water tens of metres, but modern electric and diesel pumps can increase that to thousands. However, once water is raised more than about 150 m, carbon dioxide is released from the water and corrosion of pipes and pumps is the result. Also, near the coasts the water-bearing rocks have increasingly become infiltrated with salt water, rendering them useless for agricultural purposes, poisoning crops and alkalising the soils. This is the result of increasing intensification of agriculture, particularly of vegetables grown in polythene tunnels, irrigating forage crops such as alfalfa and now even irrigating the expanding viticulture industry. For the sake of conserving its scarce water resources, Mallorca might do better to concentrate on less thirsty crops and revert more to dry-land farming.

Much of the landscape of Mallorca has been fashioned by supplies of water operating differently over many centuries. Its small-scale technologies are scattered everywhere. They have shaped the location of settlement, controlled the rate of expansion of settlements, especially tourist resorts, affected agricultural land-use, crop types and productivity, and generally helped determine the appearance of the island.

Vegetation cover – mostly about trees

A final element in the appearance of the physical landscape is its plant cover. In Mallorca it is difficult to generalise geographically about the distribution of species because over the historic period any zones of so-called dominant species have broken down to give a mosaic or patchwork appearance. The vegetation that we see today, however, seems poorly adapted to the present-day climate, which has existed for only the last 5000 years, and even less so to the climate since the Little Ice Age and to contemporary global warming. Plants have not fully adapted to this, and retreat and wither somewhat during the scorching summers, although rarely do they shed all their leaves. The two wetter seasons alluded to above ensure that in addition to the true spring season there is a re-greeening of the landscape in the autumn.

Given the mountain topography and geology to the north and west, it is not so surprising to find that much of that area is actually devoid of a considerable vegetation cover, particularly over about 600 m. There may be nowhere quite like the limestone pavements of the Burren in the west of Ireland, but large areas of seemingly bare rock are a common sight (Fig. 3. 5). Away from the cultivated fields what are more noteworthy are, firstly, the trees in the landscape and, secondly, the scrubby lowland garriga. Trees tend to be of two main types: woodland trees, but also sometimes free-standing trees, and the trees that are, or perhaps have been, part of the agricultural landscape. The 'natural' tree cover on the higher ground is made up mainly of holm oaks (*quercus ilex*) and pines, principally Aleppo pines (*pinus halepensis*). These occur primarily in the Tramuntana but are now quite widely spread on the Serra de Llevant, the Artà hills and even on the minor hills of Es Pla. What is most remarkable has been the massive increase in tree cover of pines and *encina* during the twentieth century from just over 3 million trees in 1891 to nearly 34 million trees 100 years later (Berbiela Mingot, 2010).

Historically the holm oak has been the more significant landscape feature in past writings. Only in small areas does it form true woodlands, but it has had some economic importance for building purposes, for fuel (*leña*), including making charcoal and in some periods, pannage. In the prehistoric period there is little evidence that the holm oak was widely used. Recent research suggests that from the very earliest days as agriculture spread it was rarely at the expense of such woods, which often had an additional economic purpose. As society became more sedentary in prehistoric times, the concern of the population 'focused on exploitation of the tree

Figure 3.5 Treeline in Tramuntana, above Galatzó, Calvià.

stratum of the wild olive tree within the *maquis*. For centuries it was not necessary to go to other woody resources to satisfy the needs for fuel and timber. . .' (Pique and Noguera, 2002; Waldren and Ensenyat, 2002: 299). By the Roman period the pine was the most worked wood resource. In the medieval era the holm oak provided few good-sized timbers (*madera*) for large buildings such as churches or barns or for ship building, which may explain its survival. In Mallorca at least the mountains were not 'denuded of forests' between 1650 and 1800 for shipbuilding as McNeill claims (McNeill, 1992: 148). In 1748 the navy undertook a survey of Mallorca's trees and counted – very precisely – 7,186,710 of them! Of the two main species 2. 4 million were *encinas* (holm oaks (65. 5%) and 4. 7 million pines (33. 1%) (Fig. 8. 10). The *armada* was concerned more to conserve wood than to exploit it recklessly, and a strict system of felling licences was enforced, stipulating the size and shape of tree to be removed (Gil *et al.*, 2003: 152–3). The late nineteenth century did see a more concerted attack when industrialisation and urbanisation made greater demands on Mallorca's high woodlands, but even then good management ensured that extensive over-exploitation only occasionally took place.

Today the ubiquitous tree of Mallorca is the Aleppo pine, which thanks to the decline of agriculture and its concentration on the better soil and irrigated areas, and to rural depopulation, has actually led to a considerable spread of this species, although not at the expense of the oak woods. It is found in all lowland zones, increasingly in areas abandoned by agriculture. Between 1971 and 1999 its area in the Balearic Islands as a whole doubled from 55,000 ha to 115,000 ha (Gil *et al.*, 2003: 43, 229). It spreads rapidly by the usual natural processes of seed dispersal; there are few natural seed predators. Though subject to fire damage (see below) it soon regenerates through seedlings (only about 50% of seeds germinate), often dramatically so, providing there is adequate precipitation in the first post-fire season. It grows to about 20 m and is a useful timber tree, but in Mallorca its harvesting is limited. It is little wonder it is spreading everywhere!

Finally, it is important to say something about garriga, the undershrub, often aromatic, lowland vegetation that is to be found over much of the central and coastal areas. Formerly much more widespread, it has been progressively cleared over many centuries to provide land for agricultural crops from cereals to tree crops. The human effort in this has been prodigious, but over the last fifty years it has begun to reassert itself where grazing animals and ploughs have declined in number. Today its distribution is concentrated on the lower slopes of the eastern uplands and parts of the central plain. Garriga is composed of a wide variety of plants including cistus (*cistus albidus*), juniper (*juniperus communis*), lentisk (*pistacia lentiscus*), wild olive (*olea sylvestris*) and the wild varieties of herbs such as thyme (*thymus capitatus*), rosemary (*rosemarinus officionalis*), sage (*salvia officionalis*) and fennel (*foeniculum vulgare*). Interspersed will be pines and holm oaks which are part of the mosaic too, providing they can develop their more extensive root systems in competition with the shrubs and undershrubs (Grove and Rackham, 2001: 57). For 1999 it was estimated that *ullastre* (wild olive – *olea sylvestris*) and other garriga trees numbered 9. 4 million, 21. 5% of the arboreal total (Berbiela Mingot, 2010: 553).

Fire and the landscape

In recent years the appearance of a burning landscape seems to have increased throughout Mediterranean regions. Nearly every year in the eastern Mediterranean basin in Europe, the Mediterranean coastlands of Australia and in California fires make the headlines. The causes of this are not fully understood, but two factors that seem to be playing an important part are global warming (even drier and hotter summers) and human agencies as settlement pushes more and more into desirable areas. According to official statistics 99% of the Balearic Islands is at high risk from fire – a somewhat disingenuous, even misleading, statement. Nonetheless, as with water supply, a history of the role of fire in the landscape of Mallorca would be fascinating.

The important thing to remember is that fire is a *natural* part of the Mediterranean environment; indeed, for many aspects of the working of the ecosystem it is essential. Much of the vegetation we have described is both susceptible to fire and adapted to fire. Shrubs and undershrubs often have oily leaves and stems and many trees produce carpets of leaves that are easily ignited, especially if they have been allowed to accumulate. Lightning strikes are very common in the disturbed weather conditions and are a major cause of fires; prehistorically there is ample evidence of fire without human interference. At the same time there is now some scepticism that prehistoric societies used fire to clear forest to provide farmland; fire may have killed trees, but removing the burned trunks and roots would have presented major problems. In the long historical period from the Roman era to the eighteenth century in Mallorca the records of fires are rather slim. In the modern age much more is known about the extent and incidence of fires, but their causes are still debated. Accidental and deliberate wildfires do occur, but finding the culprits has proved notoriously difficult despite the press quickly seeking to blame the dropped cigarette-end or the rapacious property developer hoping to benefit from land 'cleared' by fire. As Rackham and Moody point out, 'All over the world, fires occurred long before human history, and plants and animals became adept to resisting, evading or recovering from them' (Rackham and Moody, 1996: 116).

The charred landscape after an extensive fire is a common sight in Mallorca, but it is a temporary phenomenon. Between 1969 and 1983 there were over 1000 fires recorded on the island, covering in total more than 22,000 ha. The number of fires is increasing, but the area burned does not show a linear progression. The worst-affected areas appear to correlate with the distribution of Aleppo pines and with the rate of urbanisation, so that the lowland areas of Muro and Sta. Margalida characterise the first case and the area to the west of Palma – Calvià, Puigpunyent, Esporles and Andratx – the second (Alzina *et al.,* 1984: 92). Where the two variables come together they show the worst incidence of fires. As farmland is reduced and poorly maintained lands merge together, so the number of fires seems to increase. In the woodlands themselves, such as in the Tramuntana, the reduction in exploitation has allowed wood to accumulate as fuel over time, increasing the risk of fire. Finally, the natural spread of Aleppo pines referred to above has increased the number of fires, as it is a particularly fire-prone tree, especially if it occurs in plantations. Generally speaking the risk of fire at any one time increases with the 'rule of 30' – temperatures over 30 °C, humidity below 30% and winds in excess of 30 km h^{-1}. It is always important to distinguish between small fires (*conatus*) and large ones (*focs*). The pattern in Mallorca is for a few large ones rather than a large number of small ones.

For the landscape historian and the ecologist an important concern is the recovery of burned lands from the effects of fire; rarely is the landscape post-fire the same as that pre-fire. Regeneration is dependent upon a number of variables, including the species concerned and the environmental conditions after a fire. For example, a fire does not sterilise the land for long, and more types of plant grow from the ashes than existed before. Shrubs, such as the wild olive, tend to recover more rapidly than the smaller undershrubs of the garriga, and many types of trees find it difficult to compete with these more vigorous shrubs post-fire (Eugenio and Lloret, 2006). If the weather conditions after a fire are wet then recovery is faster. If farmers introduce a different form of agriculture into the burned area, then that can affect the success of many plants' recovery. We also know that in some of the mixed woods of Aleppo pine and holm oaks found in Mallorca, recovery nearly always benefits a single species in the long run (Broncano *et al.*, 2005: 47–56). It is difficult to generalise about post-fire vegetation, but together fire and its aftermath have been and remain significant visual and ecological elements in the island's landscape mosaic.

In conclusion, the physical environment and its dynamics that we have described here form the stage upon which human occupation, interpretation in terms of social and economic value and general exploitation have taken place over the last four and a half thousand years. The physical and vegetational landscape was certainly not immutable over the long period of settlement, indeed much of it has been altered by human agency, but the environmental context has to be taken into account when explaining the prehistoric and historic development of Mallorca's landscape. It has constrained choice in different ways at different times.

4

Prehistoric Mallorca – early human imprint

The very earliest cultural landscapes in Mallorca are those created by prehistoric societies. The natural landscape features of the island had, of course, changed many times in the post-glacial period largely as a result of climate and sea-level changes. In future, technology and human ingenuity would be much more influential than natural forces and over a much shorter time period. What begins with the establishment of temporary settlements on the island continues some four and a half thousand years later with the industrialisation of the economy, the high-rise tourist hotel, and the complexity of Palma's urban structure.

In the popular imagination the prehistory of Mallorca is often perceived as consisting only of the readily visible *talayots*, a form of megalithic building which, according to Pericot García, 'has set a stamp on the landscape which neither the passage of time nor depredation by later inhabitants has managed to efface' (Pericot García, 1972). However, these structures are a product of a *protohistoric* age; their landscape was one the Romans would have encountered in 123 BC. But long before written history begins there were three millennia of human settlement that make up the story of the *prehistoric* colonisation of Mallorca from its beginning, including many centuries of landscape change.

Who and when: the debate about early settlement

When was Mallorca first settled and by whom? These are questions which island archaeologists have been debating intensively for the last three decades. It is not possible here to enter into the complex issues involved, for our task is to examine the impact of early settlers on the landscape. However, since a difference of about three millennia exists between the claimants for an early settlement and those who argue for a later one, it is necessary at least to summarise the main points, if only because within such a large time difference considerable landscape changes may have taken place for which human agency may have been responsible.

An earlier generation of archaeologists believed islands like Mallorca to have been initially settled from other islands; Pericot García thought contact with the mainland was 'almost without consequence' (Pericot García, 1972: 111). There is a theory, derived from island biogeography, however, that postulates that settlement of an island depends on its size and configuration and on its distance from its nearest continental landmass; this is known as the *target/distance ratio* which, briefly, means that large islands near a continent are likely to be occupied before small islands at some distance (Patton, 1996; Boomert and Bright, 2007). If islands have distinctive shapes and outlines, especially high mountains, then this adds to their propensity to be 'discovered' and explored or occupied early. All this is predicated, of course, on the generally accepted idea that islands are settled after and from continents.

In the Holocene period, about 9000 BP, Mallorca was the sixth largest island in the Mediterranean with an area of about 3740 km² (after Sicily, Cyprus, Crete, Sardinia and Corsica) and it lay some 90–120 km from the coasts of Spain and France. Its target/distance ratio was therefore quite small (Fig. 4. 1) suggesting a later rather than an earlier settlement, although the height of its tallest mountain (Puig Major, 1445 m) and its square configuration might have moderated this somewhat. Some have argued that early visitors may have come via the Columbretes Islands off the coast of Valencia, using a stepping-stone approach (Waldren, 2002: 160).

The second argument concerns dating. With the advent of radiocarbon methods it is now possible to speak more accurately about the dating of human artefacts and animal remains. The Balearic Islands are fortunate to have over 750 radiocarbon datings, two thirds of them from Mallorca, the earliest of which suggest a human presence on the island from about 5600 BC (Mico Perez, 2005). This date is from the cave of Son Muleta near Valldemossa. However, early dates do not necessarily mean early permanent settlement and occupation – prehistorians like to distinguish between discovery and colonisation. Modern archaeology, linked as it is to anthropology, demands cultural evidence of settlement too. What evidence is there from artefacts for permanent settlement, or is any such evidence merely of temporary, perhaps seasonal, occupation? That early settlement of some kind was possible from at least the sixth millennium cal BC does not appear to be in doubt:

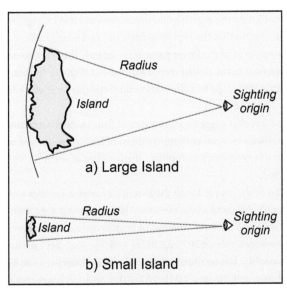

a) Large Island

Radius

Island

Sighting origin

Radius

Island

Sighting origin

b) Small Island

Figure 4.1 The target/distance ratio. The larger, higher and closer to other landfalls, the more likely an island is to be 'discovered' early by potential settlers. The consensus is that Mallorca was first settled from the Peninsula. Source: after Patton.

'The claim . . . that people were not capable of putting to sea on rafts or in boats during the Pleistocene is nothing more than an instance of temporal chauvinism' (Cherry, 1990: 201). It seems more likely that a move towards permanence may have occurred in stages: discovery and exploration; frequent visits; stable settlement or colonisation (Guerrero Ayuso, 2001: 139).

Settlement and resources

Nonetheless, many modern authorities would look for a cultural explanation based on the economic exploitation of the island's resources or on a set of cultural relationships. What would have sustained early settlers or induced them to remain? Such visitors might be described as late Mesolithic hunter-gatherers, and it is thought that the attraction in Mallorca was *myotragus balearicus*, a short-legged midget antelope-like creature whose remains have been found at a variety of sites on the island. This animal had probably lived on the island for millions of years, and one school of thought believes humans and *myotragus* co-existed from the sixth millennium cal BC (Davis, 2002: 208). Some have even suggested that *myotragus* may have been domesticated, but this is highly unlikely. On the other hand, others believe that this animal was rapidly wiped out, perhaps as quickly as within ninety years of contact with humans. This means that this meat supply would have been exhausted and there would have been one less reason for permanent settlement.

Was it possible that early settlers came for the island's wood resources? In the Holocene, at the end of the Boreal and the beginning of the Atlantic phase, as described in Chapter 3 (about the sixth millennium BC) the predominant trees of holm oak, elm and pines would have been concentrated in the Tramuntana, but would probably not have been the dense and impenetrable virgin forests beloved of some writers – *'una massa boscosa, espessa i continua'* – as Guerrero has described it (Guerrero Ayuso, 1998: 13). By the eighth millennium considerable changes in the vegetation were occurring, so that the typical garriga and savannah landscape was already widespread over most of the lower parts of the island. There is also no evidence of human interference in the pollen record from about 8000 BP to about 5000 BP, with no large-scale landscape transformation until the third millennium BC (Ramis and Alcover, 2001: 266).

Analysis of the palaeodiet also suggests an absence of fish, another attractive resource – marine and freshwater – a rather curious omission for hunter-gatherers, but less so if settlement was transitory before the third millennium BC (Van Strydonck *et al.,* 2002: 196).

This resource-based analysis, coupled with the absence of artefactual archaeological evidence, increasingly points to a late, even very late *permanent* settlement of Mallorca. To quote Ramis and Alcover, 'the age of human arrival on Mallorca is, in fact, not proved to be prior to about 2070 cal BC at >68% probability', so that Mallorca and Menorca were 'the last territories in the whole Mediterranean to be colonised by humans' (Ramis and Alcover, 2001: 267). This position has recently been reinforced by a comprehensive review of all the evidence by Mico Perez (Mico Perez, 2005). In accepting a third millennium date for the permanent settlement of Mallorca, we are also saying that any large scale landscape change before that date can be accounted for principally by natural agencies, especially by climate change (Alcover, 2004).

The Copper and early Bronze Ages

Until the mid-third millennium BC there was a limited number of small tribal groups of hunter-gatherers visiting the island, staying perhaps for seasonal periods; but, by and large, they would have lived in equilibrium within the existing ecosystems and changed them, and the island's landscape, very little. It was not until the beginnings of the Neolithic era with its metal tools, its pottery, domesticated animals and settled agriculture, accompanied by a superior social organisation, that the human impact on Mallorca's landscape became pronounced.

From about 2500–2000 BC settlement would have been sparse but permanent with cave dwelling continuing perhaps seasonally, and more sedentary living quarters being built. An example can be found at Son Ferrandell-Son Oleza on the Pla de Rei near Valldemossa (Waldren, 2002; Waldren and Ensenyat, 2002: 301–11). Here a sort of 'granja' of about 40,000 m^2 was constructed of two houses/ dwellings, supplied by a water channel with a simple walled enclosure of 3600 m^2 to corral a variety of animals including sheep, goats and pigs. This tribal grouping of about 12–16 individuals, who came from overseas, probably the Peninsula, also grew wheat and barley, using flint sickles and storage pottery. The settlement appears to have existed for about 1200 years, being abandoned by about 1300 BC, probably because of over-exploitation of the grazing resources leading to local soil erosion.

The long Bronze Age

When looked at from such a great distance over time, all prehistoric societies appear transitional. Despite their longevity, each period seems to be fading out of its predecessor and into its successor. As the Copper Age rose briefly at a few limited sites in Mallorca, so our attention is soon shifting towards the landscape and culture of the Bronze Age, from a landscape in which the human impact was very sporadic and limited, towards a more expansive era in which men and women began to challenge nature's dominance more thoroughly, thanks to new technologies, better developed forms of social organisation and above all, by sheer force of numbers.

The late Copper and the Bronze Ages saw the emergence of new forms of settlement, a move away from the caves and rock shelters of the uplands towards the Mallorcan lowlands. As the so-called Neolithic 'revolution' introduced domesticated animals and the beginnings of cultivation, and a move away from a dependence on the natural environment to provide sustenance, so it became possible to store surpluses of food, leading to an economic division of labour. Societies may have remained tribal, but not all members would have been occupied in the same activities or with the same skills; communal life may have been the norm, but within communities specialists would have emerged – potters, metalworkers, stonemasons, farmers.

The long period from Son Oleza-Son Ferrandell to the beginnings of the Classical Age, say, from about 2300 BC to the late sixth century BC, saw two cultures flourish, distinguished by their settlements and dwellings. First, the navetiform culture of the Bronze Age beginning about 1700 BC introduced the upturned boat shape of the new stone houses located mostly on the plains of Mallorca (Aramburu-Zabala and Riera Campins, 2006: 11). An exception to this rule is a cluster of such features in the uplands of the Boquer valley near Pollença. Usually there would be three to five houses (*navetes*) in each settlement, some of which may only have been used seasonally in the early stages (Guererro Ayuso, 1998: 58). These houses were 15 to 16 metres long by 3 to 4 metres wide with olive wood roof timbers supporting a brushwood covering. Later some roofs had a clay covering. Many, such as Poblat de Canyamel (Son Servera), S'Hospitalet (Manacor) and Es Closos de Can Gaià (Felanitx), were located near the coast, the latter dating from 1770 to 800 BC (Oliver Servera, 2005: 247) (Figs. 4. 2 and 4. 3). These sorts of locations were probably chosen because land clearance would have been much easier, given the savannah and garriga-like nature of the vegetation cover. Clearance of the shrubby undergrowth and fairly sparse tree cover would probably have been by fire. Unlike on the mainland, there have been few discoveries of stone, copper or flint axes for this period; the island had only scant resources of copper and most was imported (Guerrero Ayuso, 1998: 44). Such clearances enabled an intensification of production based on grazing sheep and goats. Some cereal growing would have taken place, as evidenced by the recovery of flint sickles and stone querns. However, with the introduction of bronze tools (there is no tin on Mallorca, so this also would have been imported) it was possible to attack the environment more efficiently. Within the next 500–700 years, beginning about 1750 BC, Mallorcan society and its landscape were transformed by the spread of livestock farming and its expansion into more marginal areas. The sheep/goat economy made greater demands of the environment as new pastures were carved out of the *garriga* and the low density tree cover of the plains. This more secure production of food helped encourage a growing population, leading in turn to a further reduction in the vegetal cover: '*els ramaders es converteixen en una forca destructiva de la mass boscosa en la constant recerca de nues pastures*' (Guererro Ayuso, 1998: 82: 'pastoralists perforce converted the dense woodlands in the constant search for new pastures'). In reality the denser stands of woodland would probably have been attacked last.

Goats were especially destructive and difficult to control, and pollen analysis shows an increasing depletion of tree species from about 1400 BC. Nonetheless, the landscape of this navetiform culture would most likely have been characterised by

Figure 4.2 Early Bronze Age navetes at Hospitalet (Manacor). Navetes were boat-shaped dwellings usually arranged in small tribal groups.

Figure 4.3 Navetes at Closos de Can Gaià, Portocolom (Felanitx).

a pattern of punctiform (unconnected) clearings in the garriga/savannah, perhaps with small tribal groups moving on after some years as resources became exhausted, allowing regrowth and regeneration.

In order to provide some feeling for the growing pressure on the environment, Guerrero has attempted to calculate population numbers over this period, though based on some rather questionable assumptions. If there was an average of 6 dwellings to each of his estimated 325 navetiform settlements with 40 inhabitants in each one, this would give a population of 13,000 with a density of about 5/km^2. By the end of the navetiform era, say 900 BC, the population of Mallorca would have reached about 30,000 at a density of about 8/km^2 (Aramburu-Zabala, 2002: 521).

The age of the talayot: Mallorca's equivalent of the Iron Age?

Talayots are substantial cyclopean structures mostly round, sometimes square, basically towers around which are clustered a series of dwellings to form a settlement. Some of them were later walled. The classic example is Ses Païsses (Artà) but they are readily visible throughout much of lowland Mallorca. Recycling and reuse of many of the massive stones such as appears evident at Son Pou (Vilafranca de Bonany, Fig. 4. 4) has resulted in their displacement. It seems likely that modern aerial survey techniques and landscape archaeology may reveal there were many more than the 270–280 settlements identified in the 1970s. Indeed, by the early twenty-first century more than 400 have been enumerated, giving a density of about one Talayotic settlement per 9 square kilometres.

Figure 4.4 Large flat stones probably recovered from the wall of a nearby talayot, Pou Celat, Porreres.

The question arises as to their origin. Were they the product of a new culture from about 1200 BC, perhaps via a Sea People, or did they evolve from the *navetes* that preceded them? Certainly they have some resemblance to Sardinia's *nuraghi* but the prevailing opinion today is that they are an evolutionary form:

> El pavón de asentamiento de la cultura talaiotica en Mallorca es el resultado de la evolución de las tendacions que venían gestándose durante el secundo milenio ANE, en particular crecimiento demográfica en media ínsula. (Aramburu-Zabala Higuera, 1998: 223)
>
> (The balance of opinion about Mallorca's talayotic culture is towards its evolution in the second millennium BC, particularly the increases in population.)

Mallorcan prehistorians and archaeologists now argue that the pressures of a growing population and a more secure food supply brought about a restructuring of society into chieftain-led tribal groupings which increasingly sought to occupy and define territory, resulting in a new form of spatial organisation. The tower-like element at the heart of the new settlements may have had a functional purpose as a look-out, but more probably it came to represent in symbolic form a statement of land ownership and a willingness to defend it (Fig. 4. 5). The talayotic landscape would then have appeared as a series of clearings with a gradation of 'value' of land from the talayot settlement at the territory's core towards a periphery, perhaps

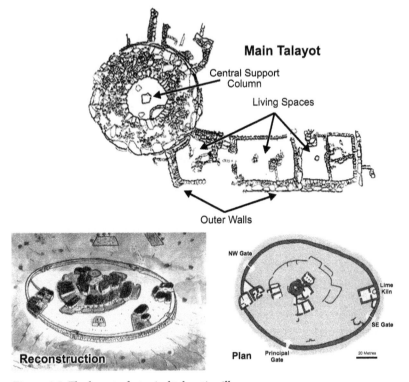

Figure 4.5 The layout of a typical talayotic village.

marked by the edge of agricultural/grazing land where it faded into the uncultivated 'bush' and a neighbouring settlement. There may have been rituals associated with this, perhaps involving demonstrations of strength at the centre and of passage into and out of a tribe's territory. Ángel Rodríguez Alcalde has argued, using a variety of locational analysis techniques, including nearest neighbour analysis and principal components analysis of environmental and hypothesised economic data, that settlement by talayots followed three phases in Mallorca: initial settlement into as many as eleven small and isolated centres, which then expanded by colonisation (the clearing and winning of land) into perhaps seven or eight zones, and finally a contraction and consolidation back into something like the areas of original settlement (Rodríguez Alcalde, 1996: 167–192). Although only detailed archaeological analysis will shed light on these possibilities, it suggests that even in this late prehistoric phase settlement exhibited the kind of dynamics we might expect to see in later historic periods.

Whereas the navetiform settlements had no public buildings or spaces or communal structures, at least until later, as at Es Closos de Can Gaià, the talayots more clearly had proto-village forms in which were embedded physical social functions (Hernandez Gasch and Aramburu-Zabala, 2005: 52). The vast majority (90%) are located between 0 m and 200 m above sea level, nearly 60% on low hills or low ridges, close to *torrents*, about 2. 5 km apart, each with a territory of about 11 to 12 km². Cereals were grown, but the expanding economy mentioned above was still dependent upon sheep and goats, particularly in the lower areas. Sheep were a source of meat, milk and wool, but goats gave a higher yield of milk but no wool. Aramburu-Zambala has produced a land-use model for each settlement, which has 477 ha for sheep and goats (50%), 211 ha for cattle (22%) and 266 ha for woodland (27. 9%). From this he has extrapolated that the 270 talayotic settlements he recorded may have controlled an island herd of about 146,000 sheep, a figure curiously close to the estimate for AD 1585 (Aramburu-Zabala Higuera, 1998: 100). By the third phase of the Talayotic era (406–123 BC) – what might be equivalent to the Iron Age – grain production appears to have increased to occupy about 6% of a talayotic village's agricultural area, especially in the 'marinas' of Campos and Llucmajor, indicative of an intensification of production, bringing further shifts in the landscape. This is now accepted as a *post*-Talayotic phase in which a small number of larger settlements, each probably dominated by a clan chieftain, held political sway over larger areas of the island in a hierarchical fashion. A good example of this is at Son Fornès (Montuïri) where recent excavations have recorded marked changes in the nature of the settlement from Talayotic to post-Talayotic up to the Roman era.

Not all of these settlements have walls, but where they do it was in this last phase that they were erected. Were they a defensive structure or a means of defining proto-urban/village spaces? In the case of Ses Païsses the wall was constructed between 540 and 450 BC (Fig. 4. 6). Its perimeter is 320 m long, it

Figure 4.6 The iconic talayot of Ses Països, Artà. Note the surrounding wall with its lintelled gateway.

is 3. 60 m high, and it has been estimated that over 8000 tons of limestone were used. Its sheer size indicates a high level of social organisation and co-operative, or maybe coerced, effort. The south-east gateway, the most famous, is orientated at 127 degrees, coinciding roughly with the summer solstice (Aramburu-Zabala and Riera Campins, 2006: 49).

To summarise, the prehistoric landscapes of Mallorca developed slowly, beginning with sporadic, perhaps seasonal, visitors and temporary settlers barely changing the natural ecosystem. With the introduction of new technologies in agriculture and metallurgy, island society started on a long road of modifying that landscape from the middle of the second millennium BC onwards, moving out from the caves and rock shelters to the open-air occupation of the more easily cleared areas of lowland garriga and savannah. By the late Bronze Age some have seen a more orderly imprint of human activity, culminating in the more rational and ritualistic landscapes of the talayotic era.

By about the beginning of the fifth century BC Mallorca was entering the Classical, historical period and being drawn more and more into the world of Carthage and Rome, and fleetingly the Phoenicians. Contact with North Africa, the Italian peninsula and its islands, possibly the Mediterranean Levant and, of course, mainland Iberia rapidly increases for trading and political purposes. Prehistoric Mallorca had never been geographically isolated since its peopling had come from abroad, but its landscapes from now on were to be open to new influences. The words 'natural' and 'landscape' could no longer be easily conjoined.

5

Roman and other empires in Mallorca:
limited landscapes

*Pirates, interferències comercials,obtenció de noves terres, utilizació dels
foners baleàrics per part de l'excercit roma, rutes de connexió rapides; totes
aquestes causes, i per ventura mes son les que s'han de tenir en compte en
justificar la conquesta de Mallorca…* (Maria Orfila, 2000)
(Pirates, the disruption of commerce, the winning of new lands, the use of Balearic
slingers as part of the Roman army, being on routes with rapid connections; all these
were used to justify the conquest of Mallorca.) [Author's free translation]

Early in 2006 one of the local banks in Mallorca, the 'Caixa', sponsored a wonderful
exhibition on the Roman period in the Balearic Islands in the beautifully restored
Gran Hotel of 1903. Its comprehensive displays and authoritative catalogue gave
Mallorcans an opportunity to see for themselves how the Roman Empire from the
first century BC to the fifth century AD had affected their homeland. Regrettably,
few seemed to take advantage of the occasion. In some ways this was not so
surprising, since the visual evidence on the ground of 500 years of occupation is
not great. There is nothing like the extensive remains to be found in the Peninsula.
There are no large scale monuments to be found, the only exception being the
recently excavated and well-preserved remains of Pollentia, the Roman city lying
to the east of modern Alcúdia. Roman Palma, a second major settlement, lies
largely undiscovered beneath the roads and pavements of the medieval core of
the present-day city. Maria Orfila, the distinguished archaeologist and historian
of classical Mallorca, rather disingenuously wrote about her fellow citizens, 'What
do the inhabitants of the Balearic Islands know about Roman times? If we raise the
question … I do not know if we would get an answer' (Orfila Pons, 2005: 178).

Despite this apparent lack of interest on behalf of the general public and
the paucity of the still visible remains of the Roman period, the evidence for
reconstructing the landscape of the time and for searching for its relict features
comes from the archaeological and written records. While the latter give us an
insight into the narrative of the Roman occupation and settlement, and has been
minutely examined by Mallorcan classical scholars (see Blanes *et al.*, 1990), it is
mainly from the archaeological evidence that we draw most of our conclusions
about landscape.

Contacts between the Talayotic culture of Mallorca and the Roman Empire predate the actual conquest and settlement. Indeed, as early as 217 BC islanders sent a plea to Scipio Africanus seeking membership of the Empire. However, it was not until the harassment of imperial shipping by Mallorca-based pirates became intolerable in the first century BC, that a decision was taken to subdue the island. Quintus Caecilius Metellus was sent to undertake the task in 122–123 BC, bringing the island within Hispanic Citerior based on Tarragona. This occupation was not fully complete until about 75 BC when 5000 *hoplites* or heavy infantry were sent to Mallorca as part of Sertorius's War. [1] After initial subjugation, a policy of settlement and colonisation was followed with 3000 Romans drawn from the mainland (Richardson, 1996: 83). This meant that land would be allocated for the *coloniae*, mostly retired Roman soldiers. It was expected that such a group would then act as an advance force likely to encourage other settlers from elsewhere in the Empire. Roman colonisers usually sought to impose what they saw as their superior culture on the 'barbarian' indigenous population – founding towns, laying out new field boundaries, clearing land, building roads – eventually leading to a more refined landscape of farms, villas and towns. In Mallorca, however, only remnants of these forms have so far come to light.

Mallorca's Roman countryside

Land colonisation was preceded by a complex surveying process known as centuration. Army engineers – *gromotores* – would set out field boundaries and service roads using standard Roman instruments that could measure distance and right angles using Pythagoras's 3–4–5 triangle to produce a familiar set of orthogonal landscape features, the two right angle co-ordinates being the *cardo* and the *decumanus*. Two axial roads or tracks (*limites*) would be laid out giving a grid square with sides of 2400 ft (730 m) to contain 100 smallholdings (*centuriae*), each one large enough to support a family. This may not have been egalitarian in terms of the distribution of land by quality, but it was efficient. This process has been studied in south-east Mallorca both in the field and using satellite imagery (Rosselló Verger 1974; Montufo, 1998). Vestigial remains of the system can be seen in the contemporary landscape of Santanyí, Ses Salines and Felanitx municipalities. Another area more recently studied is in the *raiguer* near the Cami de Sta Eugènia and between Algaida and Consell on the Cami de Muntanya (Sandez Leon and García Riaza, 2005).

This highly regulated landscape provided the economic basis for rural settlement. Farms and smallholdings must have been built, though little evidence of them has so far emerged. From Roman conquest and settlement elsewhere it would be normal to expect to find nearer the top of the hierarchy a patterning of the Mallorcan countryside with villas, either as imperial implantations or as a form adopted by the local tribal aristocracy. In Mallorca there are but few examples, and certainly nothing to compare to the intense Romano-British patterns found in

England. Maria Orfila has listed nine major sites in three groups: the first, consisting of Sa Mesquida near Santa Ponça, Can Maiol in Felanitx and Son Joan Jaume in Manacor, has the best known remains, especially at Sa Mesquida, where a building of nine rooms with a well and a cisterna has been found. The second set consists of Son Porquera in Porreres, Es Bau and Es Vela in Consell and Son Piris, Son Vives and Es Figural Blanc in Santanyí, one of the areas of intense centuration. The third group is located in the same area but associated with earlier Talayotic settlement (Orfila Pons, 1993). Another attempt to locate such a settlement has been made by Font Jaume and others at a site in Llubí and Maria de la Salut, known as Son Matet, where large amounts of pottery were found in an area without a talayotic presence, suggesting a 'de nova planta' although no actual building has been found. However, the site is located close to a number of talayotic villages and farms, which may have made for trade between the two cultures (Font Jaume *et al.*, 1995).

This reminds us that 'much of the past remained in the Roman countryside' (Dyson, 2003: 88) and Dyson suggests that new colonisers were rarely intent on, or able to, alter the rural society they conquered. The land occupied by military veterans would have been but a small part of the island. The number of new settlers appears to have been limited – isolated Mallorca was not a popular retirement destination. Thus the countryside would have remained dominated by the previous longstanding patterns of the Bronze Age and Talayotic eras, another example of both the palimpsest and the mosaic, ideas introduced earlier.

The question then arises as to whether Roman settlement and the existing Talayotic system co-existed side by side, producing two separate and distinct landscapes, or did they merge geographically? Merino and Torres, examining archaeological evidence for the Manacor area, found that over two-thirds of the finds supported the continuity of the older culture. The Talayotic occupied the coastal zone characterised by woods and *garriga*. From this it is difficult not to accept the possibility of a differentiation of economy according to the respective environments in which each culture developed (Merino Santisteban and Torres Orell, 2000).

But as we have observed elsewhere, much archaeological work remains to be done in Mallorca, and in the fullness of time many more structures and artefacts are likely to be discovered, probably from the transitional period at the end of the western Empire, as has been the case in much of North West Europe and Peninsular Spain. Until then the Roman footprint on the Mallorcan countryside seems to have been tantalisingly light.

The impact of the Roman colonisation on the vegetal landscape of Mallorca in the long Sub Atlantic period was in essence a continuation of the processes that had begun in the early Copper and Bronze ages, that is a pushing back of the woods and *garriga* of the lowlands into the more marginal areas, principally under the influence of grazing. Crops included wheat, barley and oats. Pliny observed that the wines of Mallorca were comparable with the best in Italy. Evidence from amphorae recovered off the island's coast suggests that Mallorca was not self-sufficient in olives. This kind of trade also indicates that shipbuilding was active, local woodlands being the source of timber. In the upland parts the pine woods also provided the raw material for the production of pitch used to make ships more watertight, a technique applied to the pottery vessels used for transporting oil, wine and fish sauce. Timber was the major fuel for heating (charcoal) and in industries such as lead smelting, ceramics and dye-making. However, these activities would not necessarily have had a large-scale detrimental effect on the wooded landscape, and certainly not in comparison to the agricultural effects. Despite the fact that the Romans were a colonial power they were also economically astute. It seems highly unlikely that the woods were over-exploited but rather, as in most generations, simply 'managed'. As Grove and Rackham point out (2001: 174–5), why destroy the trees upon which money grew?

The urban system

If we know little about rural Roman Mallorca, what we know most about is its urban structure, largely thanks to excavations at Pollentia, an important town in the north of the island between the bays of Pollença and Alcúdia (Doenges, 2005). The second and maybe less important city – as far as we know – was Palma. Other towns named in contemporary sources include Guium and Tucis, whose locations are unknown, although the latter is thought to have been near Petra (Sandez Leon and García Riaza, 2005: 187). One other urban settlement that probably predates Roman times and is, according to some, to be associated with the spread of Greek and perhaps Etruscan culture in the western Mediterranean is Bocchoris, near to present-day Port de Pollença. Besides being referred to by Pliny the Elder, two inscribed tablets have been found. It is suggested that a large walled Bronze Age village dating from 1000 to 800 BC existed on the slopes of Cavall Bernat above the Boquer Valley, but with a culture based on merchants and traders rather than the usual agriculture (Cerda, 2002: 12).

It might be expected that the Romans would have developed a road network to connect these urban and rural settlements, especially between Palma and Pollentia, but tantalisingly little evidence has come to light.

The site of Pollentia was first suspected in the sixteenth century, but it was not until excavations by Llabres and Isasi Ransome in the 1920s and '30s that its existence was proved. More detailed archaeological work between 1948 and 1995, supported by the Bryant Foundation of USA, began to reveal the size and extent of the city, then shown to have fluctuated in size from 15 to about 20 ha (Fig. 5. 1). It was an unwalled city of moderate size by Iberian standards, initially settled by about 1500 *colons*, probably Italians and probably brought over from the Peninsula. Tarraco (Tarragona) at 70 ha, Emerita (Merida) at 100 ha, Cordoba, 70 ha and Caesar Augusta (Zaragosa) at 50 ha were larger; Barcinum (Barcelona) was smaller at only 13 ha (Orfila Pons *et al.,* 1999). The site was laid out using the

Figure 5.1 The Roman settlement of Pollentia. Source: Estop.

Roman surveying techniques based on Pythagoras – probably 5:12:13 rather than the more usual 3:4:5. Many of the features characteristic of Roman towns are to be found here.

Religious and commercial activities for Pollentia are well documented. The theatre was one of the first sites to be brought to light (Fig. 5. 2). It dates from the first century AD, already a symbol of maturity where entertaining scenes (*ludi*) were enacted. The later discovery of a network of rectangular streets and houses in the Sa Portella area includes the House of Two Treasures, shops and workshops. Larger houses had two storeys with porticos (Fig. 5. 3). Shops often had wooden counters at their entrances with workshops for making articles such as glass and metal in the rear rooms. More than 200 necropolises are scattered around, mostly to the south of the site, near the theatre. In the 1980s and '90s the forum was exposed, whose principal structure is the Augustan Capitolium on its north side, although little of its podium remains. Commercial activities continued through most of the second century, but when pirate threats increased during the third century the town's defences were reinforced. Much of the rest of that century witnessed economic decline, and finally Pollentia was engulfed in an enormous fire which destroyed most of the buildings. However, parts must have survived, if only in a semi-derelict state, because Muslim artefacts have been excavated from amongst the ruins.

Our knowledge about the Roman city of Palma to the south is much less certain because inevitably 2000 years of urban development on the site has meant that

Figure 5.2 The theatre at Pollentia. Source: Estop.

Figure 5.3 Network of rectangular streets and houses in the Sa Portella.

archaeological excavations and discoveries have been limited. A policy of rescue work whenever new possible sites are exposed by property development is now in hand, and more has been learned in the last twenty years than in the previous 200.

The development of Roman Mallorca was slow and sporadic geographically. As we have shown, the Talayotic culture probably lasted until at least the second century BC. Recent excavations near the site of the new hospital at Ses Espases to the north of present-day Palma have unearthed a substantial settlement from about the time of the initial Roman conquest. This site, referred to in medieval documents as Alta Palma, has been proposed as the first Roman settlement in the south of Mallorca, the precursor to what was to develop as 'Palma' about sixty years later, but there is little evidence for this (Diario de Mallorca, October 22, 2010). On surer archaeological ground, in modern-day Palma it is now agreed that a small *castella* was first built on the current site of the Almudaina on cliffs overlooking the sea and the mouth of the Riera river to the west, that is, a fortified settlement above and a port-based one below. Initially the upper site would have been a rectilinear walled enclosure with a grid plan of streets, with the principal ones exiting the city via the cardinal points of the compass, with a bridge over the Riera for the west-bound road. The area of the first settlement was small at about 6.0 ha. Its forum probably lay beneath the present-day Estudio General. In 1997 the remains of what might have been a theatre were discovered, recently supported by a hypothesis that a set of existing radial alleyways supposedly forms the typical semi-circular shape of a Roman theatre. It was located overlooking the Riera between the present-day Calle Juan Carlos I and Calle Paraires near the Born (Moranta Jaume, 2007). Also, like Pollentia, Palma probably had a low-level aqueduct supplying water to the town. Beyond these few sparse findings it is difficult to go into much detail in reconstructing the urban landscape of Roman Palma. The map (Fig. 5.4), partly derived from García-Delgado Segués and Garau Alemany (2000), offers speculative insight into what things might have looked like.

A question that still puzzles archaeologists is the extent to which the built form of these Roman cities survived into subsequent eras. If they seemed alien to the culture of the succeeding Germanic peoples were they abandoned, dismantled or simply ignored and new forms of settlement begun elsewhere? We have noted the fiery destruction of Pollentia, which may have made it unattractive in any case. For Palma, any comment is mere speculation, at least until the beginning of the Muslim period (early tenth century). Then there is some evidence that these new peoples valued its site and geographical position in much the same way, with the *medina* being built on the remnants of the Roman city. In Palma it appears likely that the Muslims utilised at least the remnants of the Roman walls for the foundations of their own fortifications.

This overview of Roman Mallorca rightly suggests that the impact was limited, but maybe what we are dealing with is a paucity of finds and evidence. From a 'political' point of view of Spanish archaeology, this is a little surprising because of the support under the Franco regime for all things Roman, unlike their disdain for matters Arabic. Until thirty or forty years ago that was the case with Roman Britain, but now, thanks to aerial and geophysical surveys, we know that the English

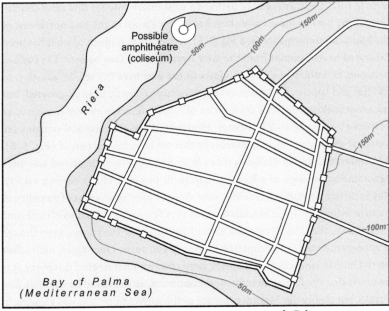

Figure 5.4 The plan of Roman Palma. Source: Ajuntament de Palma.

countryside is littered with finds from the Imperial era, even if many of them might be better described as Romano-British (Taylor, 1983: 83; Pryor, 2010: 191ff; Abulafia, 2011: 227–229). In time future research may well reveal a similar level of influence of Roman culture on the pre-existing Talayotic landscape.

Late antiquity in Mallorca: Vandals and Byzantines
Further north in Europe, the centuries succeeding Roman rule are usually known, wrongly, as the Dark Ages, but in the Mediterranean, where classical influences lasted longer, the term 'late antiquity' is a better description of the period from the fourth century. The late Roman Empire relied heavily upon North Africa for supplies of grain, and it is likely that this was a source for Mallorcan imports, especially from around Carthage ((Abulafia, 2011: 227–229). Most of our knowledge of Mallorca from about 450 AD is derived from Jewish and early Christian sources. In 455 AD an invasion by Vandals resulted in the takeover of the island by their kingdom of Mauritania, whose capital was first in Septem (Ceuta) and later Carthage in 439 AD, both in North Africa, making Mallorca once again part of a frontier between Europe and Africa (Amengual i Batle, 2003).

Christianity had probably been introduced to the island in the late Roman era, if only as a secret sect, but under the Vandals a series of basilicas (small churches with characteristic rounded apses and baptisteries) was begun, suggesting a more flourishing community likely to have been overseen by their own bishop. Four major sites of this kind have been discovered so far, all in fairly remote rural areas, suggesting perhaps a missionary role for this early church. One of the best of these early churches is Son Paretó to the east-north-east of the Manacor municipality (see Fig. 5. 5). It is likely that much of what has been excavated to date comes from the later Byzantine era (see below). The basilica occupies an attractive site with views to the east over the sea. In addition to its altar and baptistery, a number of graves have recently been discovered, but the most striking survivals are a series of mosaic floors, now well preserved in Manacor's museum in S'Enegistes. However, these ecclesiastical remains are amongst the few pieces of archaeology that we have for the period (Fig. 5. 6). The Byzantine rule of Mallorca dates from 534 AD, when the island was once again under the aegis of a Roman Empire, in this case of the Eastern variety. The inclusion of Mallorca into the 'new' Roman Empire was part of its policy of 'renovatio imperii.' If the Vandals have left very few vestigial features of their time in Mallorca, then the Byzantines under Justinian I left even fewer, even though it appears they were in occupation for nearly 200 years. Once again, their effect on the human landscape of Mallorca awaits more archaeological discovery. It is assumed that the Christian influences continued, as witnessed by sites like Son Paretó, but clearly the Muslim presence in the surrounding seas was growing in the eighth century, and raids and fear of raids saw defensive positions on commanding, defensible high points being strengthened. Examples include Castell Alaró, Santueri and possibly Castell del Rei near Pollença (Amengual i Batle and Cau Ontiveros, 2005: 198). What went on after the collapse of Byzantine interest remains a mystery, with sporadic Muslim raids a likely possibility. What is more probable is that the late-talayotic culture that predated and continued under Roman rule was still very active and becoming more sophisticated.

Figure 5.5 The palaeo-Christian church, Son Paretó (Manacor).

Figure 5.6 Tessellated pavements from Son Paretó, now in Enegistes Museum, Manacor.

The 900 years between Metellus's first conquest of Mallorca in 123 BC and the Muslim invasion of 902 AD is almost as long as the period between the Christian Catalan-Aragonese settlement in 1229 AD and the present day. And yet we know so very little about such a long stretch. Certainly the Romans and their colonists left some marks on the landscape, especially in the urban sphere, although it is only thanks to the archaeologist that these can be seen today. Succeeding imperial powers left an even smaller contribution to the palimpsest. If the second part of this long period was indeed a 'dark age', then a new enlightenment would come with a new religion and a quite new culture, from even further afield than even that of the sea peoples of the Neolithic. The Muslims and their followers were to transform the landscape of the island with towns, villages, mosques, grand estates, a remarkable facility with water engineering and a range of crops exotic to Europe, all supported by new science and new technologies and exploiting to the full Mallorca's geographical location as a pivot in the Western Mediterranean.

The landscape of the Muslims, 902–1229

In this month (September) when the level of water in the ground is
lowest, before the autumn rains begin, you should concern yourself
with the sinking of wells and give attention to all things to do with the
irrigation of fields and plots.

When seeking water in mountainous regions you may discover its
presence in a certain place when there is dampness on the surface of
the ground... sometimes you may hear a murmuring underground
which will indicate that water is close by.

(Ibn al-Awam, second half 12th century[1])

As a nation, Spain was born in the struggle against Islam, and the
Islamic element in the national culture was therefore rejected, despised
or ignored... The Islamic period... was widely felt as an unfortunate
accident, an interlude that had interrupted the normative development
from Roman times to medieval Christianity... Islam did not properly
belong to the Spanish memory. (Marin, 1998)

The Muslim settlers in Mallorca did not arrive in any numbers until nearly 200 years
after their armies had first crossed into the Iberian Peninsula from North Africa.
The reasons for this are not clear, but it was likely that the Balearic Islands were not
seen as a prize worth taking in terms of their resources and land, nor in the eighth
century did they pose a strategic threat to Arab trade in the Western Mediterranean;
or if they did, it was not worth tackling. [2] Despite their late arrival in 902 they were
to remain as lords of the islands for over 300 years and were to have a profound
and long-lasting effect on Mallorca's landscape. What they brought especially was
a sophisticated system of water engineering and management that was to help
transform the agriculturally marginal lands of the island, a system of tapping water
supplies, distribution and storage that was inherited by their Christian successors.
Their effects in this respect were much greater than the Roman settlers had had.
Evidence of their impact can be seen in the landscape today. They also brought
to Mallorca, as well as the Peninsula, a new language that has partially survived
into modern Mallorquin, and in morphological terms litters the place-names of
villages, farms and physical features, sometimes helpful in unravelling patterns of
settlement and land-use.

Muslim armies had first entered peninsular Spain in 711 AD, according to some a moment of good timing because the famine of 707–9 had halved the Iberian population (Ruggles, 2002: 4). By 721 they had crossed the Pyrenees. Where had these new conquerors and subsequent settlers come from? The popular view that a Mohammed-inspired army had swept all before it from the Middle East to the borders of France is something of an exaggeration. As a religion, Islam had its origins in Arabia in the seventh century. Like the slightly earlier Christianity, it was a proselytising religion, often intent on converting non-believers. Unsurprisingly it soon came into conflict with the Byzantine Empire to the north and the Persian Empire to the east, large areas of which had been Roman for 600–700 years until the former were defeated at the Battle of Yarmuk in 635. After Damascus and Jerusalem had fallen, the drive into northern Egypt saw the winning of Alexandria in 642. It is often thought that the route to Spain was then the swift and dramatic conquest of the whole of North Africa until the crossing of the Straits of Gibraltar in 711. The movement west was certainly swift, largely because the initial Arab forces only conquered selected areas, principally urban places strung out along the southern Mediterranean coast. This 3000 mile journey crossed a variety of types of country including desert and mountain, as well as the fertile coastal strip. It makes sense to see this line 'in terms of a small number of discrete areas of relatively dense settlement and population, separated from one another by much larger stretches of marginal land' (Collins, 2004: 121). And the rapid conquest was far from being solely land-based. With the capture of Alexandria came the acquisition of 'the largest concentration of shipping in the Mediterranean' (Collins, 2004: 122). This gave the Arabs the maritime resources to service their conquests. Carthage, which had had a Roman presence for 800 years, was captured in 698, giving them another naval base and permitting the first sea raids on the Balearics early in the eighth century. Tangiers was taken by sea in the first decade of the eighth century.

By this time the invaders were no longer solely Arab, nor even completely Muslim, because they had been joined by a confederacy of Berber tribes who provided a large number of slaves and mercenaries to their armies (Collins, 2004: 124). Nor were the armies that moved rapidly through the Peninsula very large. It suited early Spanish or Gothic writers to talk of somewhere between 7000 and 10,000 men, but the reality was probably much less than that, perhaps even as small as 2000. The opposing Gothic forces were also relatively small. 'No more than a dozen families controlled many of the economic assets of the Kingdom ... a small military aristocracy perched on top of a large civilian subject population,'[3] perhaps 20,000 Goths and a million Hispano-Romans.

The settlement of the Arab/Berber culture was not the result of this first 'invasion' alone. The important point is that throughout the eight centuries that Spain came under the sway of Islam there was an ongoing series of waves of peoples coming into the peninsula via North Africa. At various times quite different tribal groups from quite different geographical origins came either as new 'invaders' or were called up as mercenary troops to support the struggles of an existing regime. Sometimes such armies stayed for many decades and took up a quasi- permanent presence, at other times they crossed the Straits of Gibraltar, carried out their military activities, collected their booty and then returned to their homelands. The southern and eastern parts of Iberia settled by both these peoples – Arabs and Berbers – were known as al-Andalus, and it was later to include the Balearic Islands. For some invaders it was seen as a land to be colonised and permanently settled, creating a new cultural geography with its attendant 'Arabic' characteristics, but for others the sojourn was temporary or intermittent and left little legacy. Mallorca was largely settled by peoples from North Africa, the Almoravids and the Almohads. The overall effect, of course, may have been long-lasting, but their impact on the landscape of Spain was geographically varied, so that while modern-day Andalucía and Valencia bear the marked imprint of Islam, small offshore islands like Mallorca show more limited traces and relict features.

While most authorities will list the familiar landscape elements of settlement – including important cities – irrigation and hydraulic engineering, field patterns, especially terracing of slopes, and place-name evidence, it is useful to consider whether there was much variation in the type of cultural transfer, given the geographical origins of the settlers. Was there a kind of cultural hegemony across the Arab/Berber world that manifested itself in landscape terms? The popular conceptions associated with, say, Cordoba or the Alhambra, are strongly linked with the regimes that had their origins in Damascus or Baghdad, but where did the castle/village forms so prominent in Valencia come from? And what of the Almoravids of the Maghreb who came from the 'trackless wastes which lie between the most southerly cultivated areas of Morocco and the settled plains around the Senegal and Niger rivers where ... the landscape is extremely inhospitable, endless plains of gravel and sand, scorching and windswept with a few widely scattered wells and, discernable only to the experienced eye, patches of sparse grazing' (Kennedy, 1996: 154)? They were said to be a nomadic people ignorant of growing crops – 'they knew nothing about tilling the land ... nor do they know bread. Their wealth consists only of herds and their food of meat and milk' (Al Bakari, an andalusí geographer – quoted by Kennedy, 1996: 154). It was the Almoravids who settled Mallorca; the later Almohads were to leave them alone for most of the period to 1229 providing they behaved themselves and paid the correct tributes on time. The Almoravids were composed of a variety of tribes, bound together by tribal solidarity rather than religion. They had only converted to Islam in the tenth century. Essentially they were an army of conquest intent on winning *ghanīma* (booty) rather than land, coming to Spain in 1086 at the invitation of the 'Arabic' taifa kings. There appears to be little evidence that they wished to settle. A branch of the tribe, the Banu Ghaniya, was made governors of Mallorca in 1126, where they were allowed to rule even after the Almohad conquest. They made trading relations with Aragon and Genoa but relied on booty captured at sea to pay off their Almohad masters. They also had ambitions in the Maghreb which led to the return of Almohad control of the island in 1203 (Kennedy, 1996: 250). The Almohads, a Berber tribe who originated in the Atlas Mountains, were more orthodox Muslims with a background in agriculture. While their influence on the landscape of Al-Andalus was considerable, it was much less so on Mallorca, with the exception of the development of the Medina.

So were the new settlers able to make a new addition to the palimpsest, and in so doing did they in any way eradicate what came before? Who were the conquerors and the subsequent settlers? What kinds of numbers were involved? Where did they build their farms, villages and towns? Are even these north European terms for features of settlement actually appropriate? Were they an urban people or largely rural? Were they colonial exploiters anxious to strip Mallorca of its assets, or was their occupation of the island one of peaceful settlement? Above all, what was their broad environmental impact on Mallorca's landscape?

If we were writing about more northerly parts of West Europe at about this time we would be dealing with a landscape long settled with villages and farms and patterned with Germanic, Anglo Saxon and Viking towns linked by a fairly dense pattern of trackways. The woodland cover of the post-glacial period would have undergone at least its second clearance as the frontiers of agriculture expanded under waves of immigrants and the impact of new technologies, especially those associated with arable farming, particularly ploughing. At the time of the Norman Conquest (1066, only about 160 years after the Arab invasion of Mallorca), the British Isles were already exhibiting features of nearly 2000 years of human change to the physical landscape. Mallorca, at about the same time, was much less of an 'ancient land'. Some woodland clearances had taken place under the influence of the new metal cultures of copper and bronze, but settlement was sporadic and population numbers and densities quite low, although, as we saw in a previous chapter, perhaps greater than originally thought.

The Arabs and Berbers who came over inherited some landscape features, but the extent of human impact from the first century BC to the tenth had not been particularly great. There were the Roman towns of Pollentia (Alcúdia) and Palma, there is the evidence of centuration described previously, and there was probably some villa settlement, but little evidence that island people adopted Roman ways. There is evidence of Roman water supply engineering, and some Visigoth patterning of the ground, but population numbers and densities were so small by the early 900s that the new settlers could largely, through force of arms if required, take up what lands they wished and build upon the few late classical foundations. What still figured most strongly were the remains of the late Talayotic 'villages' and the remnants of Late Bronze age farming. [4] While the mountain areas to the north and west were heavily wooded, much of the rest of the island showed a fairly 'open' pattern of vegetation.

The 'Arab' settlers came originally under the sway of the Cordoba caliphate and perhaps brought with them the core features of Moorish culture, including certain features of agriculture and settlement. By the time the tafia of Denia took over control, these were well established. The Almovarids and Almohads brought different traditions, especially those associated with irrigation and ter-racing of difficult slopes. Above all, the settlement was tribally based, many of the original family associations being captured in the Arabic place-names that have come down to us in Catalan or Castilian forms today. Mallorca, it should be remembered, was, particularly in the early phases of 'Arab' domination, very much an extension of al-Andalus. Just as Mallorca was later to become part of 'Spain' so the island was at this early date part of a larger political and geographical reality. At this distance and from the evidence available, it is difficult to speak of indigenous Arab forms unique to Mallorca. A more pertinent question is: to what extent was Mallorca an extension of al-Andalus in landscape terms? The answer may lie in an examination of a number of landscape elements.

Water supply and irrigation
What was crucial to both settlement and agriculture was water supply; indeed, it would not be exaggerating to say that water is the major cultural determinant of the geography of the Arab and Berber period in Mallorca and its associated landscapes at this time, although there is a considerable debate as to its origins (Barceló, 1996a). Water management is also a good example of technology acting as a mediator between society and its environment, a process referred to in an earlier chapter (Glick and Kirchner, 2000). There has long been a debate on the true origins of the water management systems in Spain: did the Arabs inherit them from the Romans or were they newly imported (Butzer et al., 1985)? Much appears to depend on what happened between the decline of Roman power after the fifth century and the Muslim invasions. In the case of Mallorca we have shown that the impact of Roman and immediately successive cultures on the island were limited, and it does not seem unreasonable to assume that most of the systems were new introductions.

Settlement was largely in those valleys where the supply of water was most likely to be available and predictable. The Berbers, especially, introduced to Mallorca their particular form of hydraulic engineering from North Africa, although it had its origins in the Middle East. It consisted of three elements, the *qanat*, the *sèquia* and a variety of storage tanks. The first of these was a complex system of tunnels driven into the hillside to tap sources of running water, usually springs which could then be utilised when water table levels fell during the dry season. Often these tunnels went some distance underground, and the longer ones would have had a series of vertical shafts going down into them to permit access for cleaning. [5] Where the water then surfaced it would be led away down slope in the *sèquia* or channel. Sèquias were usually roughly parallel to the natural

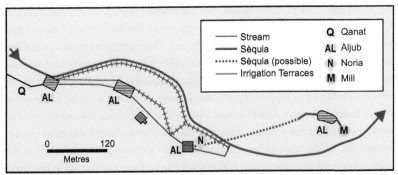

Figure 6.1 A typical irrigation system dating from the Muslim era. Source: after Barceló *et al.*

Figure 6.2 The principal *sèquia* from the Coanegra irrigation system undergoing restoration.

valley, which would have been dry from at least April to October, and often slightly above it (see Figs 6. 1 and 6. 2). Water would then be stored in one of a number of large storage covered containers such as *aljubs* and *cisterns*. The former were often located on the main driving routes for sheep transhumance from the south-east of the island to Tramuntana, linking the Muslim farms (*alquerias*) such as those in the present-day municipalities of Llucmajor and Campos to Senelles, Santa Eugènia, Santa Maria, Alaró and Bunyola. Cisterns had a somewhat different function used to store rainwater and well or running water, usually for public use in a settlement. In cultivated areas water from the sèquias was usually stored in an open *safareig* (Arabic, *s'harij*). They had to be strongly built to withstand the pressure and weight of the water they contained. The water was then led via smaller channels to a network of small fields that made up the *horta*, the distinctive irrigated landscape that surrounded a village or large estate. Here distribution was controlled not only by a series of gates, but also by complex laws that controlled how much, and when, water could be used by a particular alqueria. The size of each irrigated area was designed to support a particular population; when population numbers rose, either new irrigated areas would have to be added or the intensity of cultivation raised to feed the additional people. This kind of landscape was most evident around the Medina Mayurqa (Palma) where its gardens provided much of the city's fresh food. However, such features were to be found around all of the larger settlements, especially if located where two valleys met or where a string of open spaces in a valley could be linked together.

Examples of these landscape elements have been studied in detail by Helen Kirchner on what were then heavily wooded south-facing slopes of the Tramuntana, and relict features of them can be found today (Kirchner, 1997). One example is the l'alqueria d'Alfabia, where a small tributary *torrent* to the main Bunyola valley is served by Font d'Alfabia with a qanat 16 metres long (Kirchner, 1997: 2005). Another example is l'alqueria de Benisaba to the north of the Son Palou estate, served by a spring – font de Sa Mata or font Vermella, named after the yellow colour of the land here. Contemporary Muslim data are hard to come by, but a later document of 1256 gives production equivalent of 829 litres of olive oil from 414 trees on 2. 9 ha, by that date a typical Mallorcan *horta* enclosed by a wall and with irrigated terraces, located below Benibasa (Kirchner, 1997: 219). The question arises as to why the density and spacing of settlement varies from valley to valley, and why the irrigated areas differ; for example, that near Bunyola cited above was 10. 2 ha in extent, but that near Alaró, 30 ha.

The triangular valley of Alaró has also been studied in some detail (Riera Frau and Soberals Sagreras, 1992: 61–73). Its development was almost certainly Berber. It allowed for the upper part of the valley to be irrigated and provided a series of locations for water mills lower down. On the hillsides there were terraced and irrigated fields with water drawn from two sources, La Bastida and Les Artigues. The latter is a qanat in the valley bottom, while the former is today covered by a water bottling plant. Their outflow served four sets of alquerias or hamlets (for settlement, see below) including a first set made up of Inmalasen, Abeu Nasser, Aben Abdala, Abeu Ubeyed, Dalla and Pusella to the SSE of the present town; a second set is made up of the alquerias Banyols and Benauguir. Banyols is present day Son Forteza to the east of town – also thought to be a mosque site. Beniaguir is more difficult to locate. A third set made up of Beniarri, further up the valley, the site of the qanat of Sa Bastida and finally, a fourth set made up of Atzenet to the north-north-east of town occupying the actual Puig de Sa Comuna above the sèquia from the Font des Artigues (Fig. 6. 3) Today it is difficult to trace the perimeter of the original irrigated areas of the Berber alquerias, since the whole valley system was much altered when taken over by the Count de Béarn, part of his reward for his role in the island's conquest in 1229 and let by him to a variety of tenants. The mills – there may have been as many as nine – were strung out along the valley above the present town, but modern urban development had obliterated most of them or converted them to other uses (Fig. 6. 4).

In the lower-lying areas of Mallorca the Arab/Berber settlers developed more conventional wells for water supply, digging down into the underlying water table

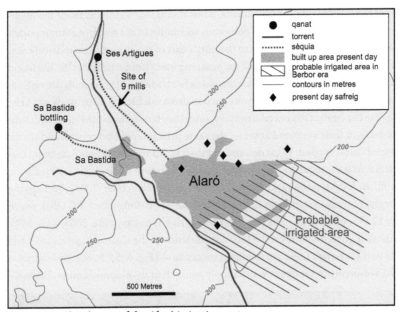

Figure 6.3 Sketch map of the Alaró irrigation system.

Figure 6.4 Alaró today showing the upper part of the irrigation system dating from the Muslim era but adopted and adapted during the medieval period by new Christian settlers.

whose depth tended to fluctuate according to the season. Therefore, storage of well water and rain water was important, and special tanks were built to accommodate this, including the *cisterna* already mentioned and the *bassa*, the latter widely used along main routeways, often preserved in place names. The basic engineering associated with these wells would have been the *noria*, which consisted of a vertical well cut into the aquifer, into which a set of earthenware jars would be dropped on a continuous loop now known as a rosary, the raising of which would be driven by animal power (see Fig 3. 4). In Mallorca there is little evidence that windmills were used to raise water in the Muslim era. Most traditional windmills are of the medieval or early modern eras. The numerous windmills in the south of contemporary Mallorca, seen dotting the area to the north-east of the airport, are a relatively new invention and introduction, used for pumping water from artesian wells. The island has, on the other hand, a much longer and richer tradition of water mills dating from at least the Islamic period that were used to drive machinery. Their study has to be seen in the context of general irrigation, since they tended to use the same hydraulic systems, at least in upland areas, so the same principles of clan control and spatial organisation applied. What determined their location was the ability to build barriers across the sèquia's long profile to enable a sufficient vertical fall of water for the mill, the local type being known as the *torre del cup* (Kirchner, 2005). Perhaps the largest of these systems was that served by the 230 m long qanat de l'Ullal, which by 1229 supported 32 mills (Gorrias i Duran and Terradas i Jofre, 2005: 313–367). Another good example can be found in the lands of the Galatzó alqueria in Calvià district (although these are medieval successors) (Fig. 6. 5). In the lowland areas the seasonally intermittent water supply made their development more difficult.

Figure 6.5 Galatzó water mills, possibly Muslim in origin but certainly medieval.

Another factor that helped determine location was proximity to the major concentrations of population and markets. For example, Medina Mayurqa (Palma) was fed by three torrentes and three sèquias, and 47 of the 62 mills mentioned in the Repartiment were found here, especially on sèquia de Canet. Outside of the Medina Mayurqa, the area of Inca (in Arabic, *yuz d'Inkan*) also had a considerable number (Carbonero Gamundi, 1992: 102). Under Berber control the majority of these water mills were used for textiles and grinding stone; the conversion to grain milling came with the bread-eating Christians after 1229. As Gamundi has written, *'hi ha raons per considerar els molins d'aiga com una peca fonamental de l'organització de la societat agraria tradicional de l'illa...'* (Carbonero Gamundi, 1992: 103). ('There is good reason to see the water mill as a fundamental unit in the island's traditional agrarian society.') It was the beginning of a longstanding feature of the Mallorcan landscape.

Settlement

Settlement, while being tribally based, was organised by clan. Different forms from those on the mainland developed largely because of the sparsity of Moorish castles on the island. Research on Arab settlement of the Peninsula has shown that the key to settlement was the building of castles (Italian *castellamento*), a process which Toubert has described as 'the structures which contained habitation and organised landholding' (Toubert, 1998). They acted as symbols of the invaders' power as well as places of protection and retreat when under attack. Such castles, known as *hisn* in Arabic (plural *hūsun*) were not feudal but the focus of tribal control around which a series of dependent settlements, usually small hamlets – *qarya* in Arabic, *alquerías* in Spanish, *alquerias* in Catalan – developed (Glick, 1995: 13–17). Such alquerias would be surrounded by irrigated fields. In Mallorca, however, only a small number of Arabic castles were built compared to the hundreds in Valencia, probably the result of less need to defend clan territory internally and the relative absence of external attack, at least away from the coast (Fig. 6. 6). The island also developed a different form of social organisation from much of the mainland, especially when under the Almoravids and Almohads. The reason may lie in the extensive form of pastoral farming that developed on the island rather than the more intensive arable and tree-based cultivation – what has been called *'un pastoralisme semi-salvtage, aprofitant pastures de garriga i aigua moll'* (Barceló Crespí and Kirchner, 2000: 252). In the more mountainous area this was not the case, as Helena Kirchner has shown (Kirchner, 1997: 337). Settlements here were often more isolated, consisting of single farms served by a complex pattern of valley irrigation. As we shall see, water supply was a determining influence on settlement location, both rural and urban, but it must be remembered that this technical activity was a product of social organisation, in this case the tribal society of Arabs and Berbers. The hisn/qarya complex was present throughout all of al-Andalus, but more exceptional in Mallorca (Glick, 1995: 21). To each

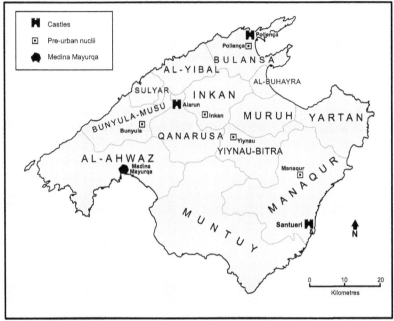

Figure 6.6 Map showing the administrative districts and major centres in the Muslim era. Source: modified from *Atlas de les Illes Balears*, 1979.

hisn there would be between seven and ten alquerias, but demonstrating the size and layout of the latter in Mallorca is not easy, particularly in the dry areas. The average size seems to have been nearly 84 ha (about 210 acres) (Poveda Sanchez, 1980), though their boundaries are now hard to define. They were held collectively, usually nucleated but consisting of many small settlements. They made up just under two-thirds of all the agricultural units, *rahals* (Catalan *rafals*), a form of private estate with an average size of just under 50 ha, accounting for the remainder (Glick, 1995: 133). So settlement was still largely based on clan organisation, but without the *incastellamento* process of the mainland.

The present-day villages were not necessarily 'villages' at the time of Almoravid development. Some may have been nucleations of *qarya* (alquerias), others isolated rafals. Identifying material features in these rural structures that date from Muslim times is not easy; the invading Christians of the thirteenth century were often intent on eradicating all signs of the previous 'heathen' occupation. Where buildings could be adapted, however, there is a better chance of finding such features. Interestingly, to an outsider, archaeological research, at least until recently, has made little effort to exhume Arab/Berber evidence, the Roman and Christian medieval periods being favoured. This means our knowledge of Mallorca from the tenth to the thirteenth centuries is much less complete than perhaps it deserves to be. Whereas the local museum in Manacor (Enagistes) has marvellous examples of Visigothic and palaeo-Christian finds, it contains relatively little from the Muslim era, and yet its impact was so much greater.

The siting of Arab settlement was affected by three or more factors. Locations away from vulnerable coasts remained important for centuries; fear of attack by corsairs was a very real threat to Mallorcan settlements for 1000 years, from the pre-Roman era to the seventeenth century. Relief, giving a defensible position, was particularly important in a land where the *hisn* was scarce. Elevated sites were favoured; even small rises of land in Es Pla, for example, were significant in affecting Arab settlement. The siting of wells and water storage facilities for human and animal consumption and for crop irrigation was a crucial determinant of settlement, and was to remain so. To the Berber settler running water was of particular cultural significance, and much effort in civil engineering terms was expended in bringing streams to the more important estates, for display and recreational purposes as much as utilitarian ones. A fourth factor would have been the degree and type of social (clan and tribal) spatial organisation. Attempts have been made to fit Central Place Theory to the distribution of Arab farms and villages in Mallorca with a view, it is assumed, to see if any kind of rational order attended settlement patterns. The hexagonal patterning seems pushed to its limits, however. More significant was the way in which tribal councils divided up the occupied land.

There were thirteen Arabic/Berber *juz* or administrative districts (see Fig. 6. 6) of which the largest, *Manqūr*, is a good example of the occupation of an area of low hills and valleys in contrast to the better known Tramuntana uplands. Its principal settlement (present-day Manacor) was located at the junction of many routes, serving as a market for the surrounding rural area of 173 scattered *alquerias* and *rafals* whose economy was largely devoted to rearing sheep and goats. Two mosques served this large rural community.

Place name evidence

Mallorcan geographers and historians have laid great store by the use of place name evidence[6] to examine the Arab/Berber origins of the settlement patterns. The *beni-* and the *al-* prefixes are often a starting point. Examples of present-day villages include Algaida, Alcúdia, and Binissalem. Later Catalan settlers naturally imposed their cultural identity on many places with their own names. This can lead to confusion so that, for example, Manacor, which even has as its coat of arms, a heart in a hand (French – *main à coeur*), is in reality derived from the Arabic *manqūr*. Indeed, the geography of place names is something of a minefield; the naming of places and the persistence of those names is largely a product of political culture. One only has to look at the renaming of many streets in Palma post-Franco to see the process at work in our own age. Nonetheless, even though the subsequent Christian settlers tried, and partly succeeded, in imposing a series of Catalan names, perhaps because both Catalan and Castilian absorbed so much Arabic, the place names of Mallorca are littered with Arabic origins. Their geography can give us a clue to what Thomas Glick has called 'the Berberisation of the countryside' (Glick, 1995: 33). Poveda Sanchez has identified eleven Berber tribal names and four Arab ones. Many of the same clan names appear in Valencia and Mallorca. From place name evidence it seems that the settlement of the island, especially in its interior, away from the Medina Mayurqa (present-day Palma), was undertaken by Berbers from North Africa, 85% of all tribal names in Mallorca being of that origin, 20% of them with Beni prefixes (Poveda Sanchez, 1980). For example, the origin of the modern town of Felanitx lies in the alqueria of the clan of Bani Furanik, a clan also found in Valencia (Barceló Crespí and Kirchner, 2000: 355).

Agriculture

When the Muslims first occupied the island in the tenth century, under the Caliphate, the principal agricultural activity was in extensive livestock farming, apparently continuing similar practices adopted in the late classical period. The sources for this information are the writings of later Arab travellers and geographers such as al-Idrisi (1100?–1165?) and al-Zuhri (died between 1154 and 1161); unfortunately there are few contemporary Arab accounts, but those most readily accessible have been assembled by the Vinases (Vinas, 2004, 186–205). In this early phase up to the beginning of Almohad rule in the early thirteenth century, the rearing of mules and sheep predominated. The mule (a cross between a horse and a donkey) was an efficient work animal, particularly for draught purposes, for carrying heavy loads, threshing and drawing water. It was a relatively docile animal, but with considerable strength. Mallorca, according to Ibn Haqwal (died after 977 AD), became a major exporter of these animals to other parts of the Arab world.

Sheep farming at this time was mainly for wool rather than meat, although the latter was clearly important to a non-pig eating society. Ibn Haqwal wrote that sheep were grazed extensively, especially in the mountain areas, and without the need for shepherds. But human population numbers remained small in this early phase, and Medina Mayurqa barely figures in contemporary Arab writings before the twelfth century. It is possible, then, that much food was imported, a persistent characteristic of the Mallorcan economy. The environmental impact of this grazing economy was, of course, detrimental to the 'natural' vegetation. Tree cover was further reduced, and with the introduction of goats, even the garriga came under pressure.

The Muslim settlers did not favour the large latifundia-like farms developed by their Roman and Visigothic predecessors, and they brought to Mallorca a 'revolutionary social transformation through changed ownership of land', basically a form of sharecropping. This, combined with the introduction of new forms of irrigation and new crops, has been called the Muslim agricultural revolution (Idrisi, 2005).

Popular mythology often quotes the Arab settlers of Iberia as introducing many new crops to Europe. In reality, of all the food crops grown in Spain only a small proportion are of Arab origin. Many of the new crops were in fact introduced by the Romans, but some came into Spain with the Arabs via India. This 'agricultural revolution … had very far-reaching consequences, affecting not only agricultural production and incomes but also population levels, urban growth, the distribution of the labour force, cooking and diet, clothing and other spheres of life too numerous to mention (Watson, 1974: 8)'. The crops studied by Watson include rice, sorghum, hard wheat, sugar cane, water melons, egg plants, spinach, artichokes, colocasia (something like a potato), sour oranges, lemons, limes, bananas, plantains, mangos and coconut palms, though not all of these found their way into Mallorca. These crops were introduced and adapted by a people 'not commonly thought to have green thumbs' and certainly not the early Almovarids from the desert areas of North Africa.

As population numbers of Arabs and Berbers in Mallorca increased through the tenth, eleventh and twelfth centuries, so new forms and types of agriculture developed. New tree crops were expanded. Al-Zuhri reported that the cultivated olive was not known in the island before Arab settlement; the carob was said to have been introduced from Carthage in North Africa. Vines were grown, but mainly for their grapes and dried fruit, not necessarily as a source of alcohol. [7] Under the Almohads (from 1203) this emphasis on tree crops increased, and more marginal land was taken into cultivation in the form of terraces on steep slopes (Fig. 6. 7). With the more widespread adoption of the hydraulic techniques we have described, the area of irrigated land increased, so that other crops such as vegetables, cotton and probably flax were grown. Grain crops such as wheat also increased, but this was much less important than further north in Europe, for two reasons: bread did

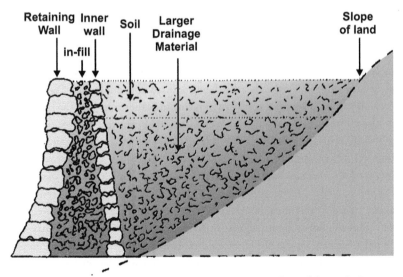

Figure 6.7 Cross section of a typical terrace. Source: adapted from Carbonero Gamundi.

not figure as widely in the Muslim diet, and there was a quite different tax system in which grain was rarely an item, both of which were to change following the Christian invasion and conquest in 1229.

We have shown that rural settlement was in the form of dependent hamlets and villages (alquerias) and rafals (private estates). How agriculture was organised is less well known. The clan-based system was largely communal and it would appear that the common features included the working of land, sharing production, and the detailed management of water supply. On the mainland the new systems were more capital intensive, especially in early years, and organisationally more complex, for example, on the legally defined allocation of water resources, while in Mallorca demand would initially have been small because of the small population and the relatively high cost of exporting overseas. This may help to explain the initial emphasis on pastoralism in the island; in truth we know little of 'what the typical (or indeed any) agricultural undertaking was like' (Watson, 1974). Land holdings varied from huge estates, through peasant proprietorships and their tenants, to the garden areas around Medina Mayurqa. The larger estates were operated efficiently by both owners and their tenants. There was little to correspond to Northern Europe's open field systems. There was no common land, and labour was supplied by share croppers – 'agricultural slaves, serfs and tenants bound to the soil . . . were rarely found' (Watson, 1974: 30).

Perhaps the best-known relict features of the agricultural system of the Muslim period are the terraces (Fig. 6. 8). However, caution should be exercised in allocating all such features to this period, as many terraced areas had their origins in the medieval and later periods in Mallorca. In upland areas such as the slopes of the Tramuntana, terracing is a means of improving soil conservation and a more rational use of scarce water resources. The actual origins of terraces is debatable. Some authorities see them coming from the Far East, others from the cradle of Arabic cultures in the Middle East. Grove and Rackham say 'written records are useless' (Grove and Rackham, 2001: 113). It is more likely that these systems were developed independently rather than being diffused from one source. Certainly the Muslim period saw a rapid spread of terracing in Mallorca, part of the revolution in agriculture of the eleventh century and later. Their construction required communal effort to raise retaining walls, develop fairly level soil surfaces, control irrigation and plant crops, including trees. The Berber clan/tribe system provided the necessary social organisation for the enormous physical transformation of slopes to be achieved; all would benefit. In the later medieval eras only slave-like serf labour could expand or maintain them. Few records remain of what kind of agricultural landscape developed on these new surfaces under the Muslims. Today they are a largely neglected aspect of Mallorca's rural landscape, most poorly maintained, many collapsed beyond recognition. However, they have become something of a resource for the 'tourist gaze' as the air-conditioned coaches drive through the mountains to Lluc and Deià.

Figure 6.8 The terraces of Banyalbufar, still in active use today, a landscape survival from the 11th century. Source: Estop.

Medina Mayurqa

The Almoravids became essentially urban dwellers, or their leaders did, and the need to develop a centre for political and religious control became important to them. It should be remembered that they had built the great Berber city of Marrakech from scratch. We have shown that the only Roman towns on the island for which there is sufficient archaeological evidence were Pollentia in the north and Palma in the south. What remained of either of them by the tenth century is far from clear. In Northern and Western Europe we know that many urban settlements of Roman origin wasted away and later settlers imposed other values on their sites. In Mallorca, Pollentia had been largely destroyed by fire at the end of the third century, but what of Palma? We now know that it is likely that it came under the sway of the Byzantine Empire and that the Roman fortified settlement was occupied in part at least by Urbs Vetus, which probably built upon the original Roman foundations; the walls may have remained largely intact. [8]

Our knowledge of the Muslim city is limited, as there has been little opportunity for detailed archaeological investigation of the site, but much speculation as to what might have greeted the new settlers. Indeed, it would be true to say, at this point in time, that 'the hypothesis rules'! (Can Ontiveros, 2004). Until archaeological evidence contradicts the idea, it would appear that Pollentia was the major urban settlement on Roman Mallorca. Therefore, why did the early Muslim settlers select Palma as their island capital? (Fig. 6. 9.) Did its site and its harbour have higher value than the northern bays? It is unlikely that all of the Roman town would have

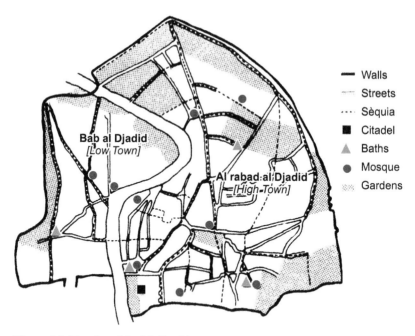

Figure 6.9 The plan of the Medina Mayurqa.

disappeared, and nineteenth century writers were convinced that the Arabs at least inherited the Roman street plan of Palma and probably the walls, most of it lying under the present day Almudaina. Roman Palma was probably deserted by the end of the fourth century AD, but the Muslim lords appreciated this area for its defensible site, on a relatively high spot above the bay and the mouth of the Riera, at which they could also build a port with a sea gate. Although founded by the Arabs at the beginning of the tenth century when Mallorca was still under the Caliphate, it was not until the establishment of one of the Taifa States based on Denia that it began to flourish.

It is perhaps difficult for the modern observer only familiar with the form and function of west European cities to understand the layout of a typical Middle Eastern town, for that is what Medina Mayurqa (Palma, Ciutat de Mallorca) was for 300 years. Our principal sources for reconstructing the plan and economic activities of the city are, curiously, not Arabic – they are rather sparse and mainly written after the Muslims had been expelled – but two medieval Latin texts. The first is the *Liber Maiolichinus de gestis pisanorum illustribus* which was written after the original Medina was attacked and much of it laid waste by a joint Pisan-Catalan force in 1113–15 AD. The second are the books of the Repartiment mentioned above and written soon after the Christian conquest in 1229. Neither is probably entirely accurate, as they are the product of forces opposed to the Muslim presence in the island. However, the latter gives a detailed picture of much of the city because this was needed by the new owners, one of whose objects was to establish themselves as major traders in the west Mediterranean, usurping that function from the defeated Muslims.

After the collapse of the Taifa state the control of the island passed to the Almoravid clan, the Banu Ganiya, which marks the beginning of a more recognisable Muslim city. The main features of the Medina would have been the original defensive core focused on the Almudaina with its own set of walls and gates, around which was built a substantial walled city, divided into two by the watercourse, an important physical feature. [9] Five bridges were built across the river to link the 'high' town (Al-rabad al-Djadid) to the 'low' town (Bab al Djadid) (Figs 6. 9 and 7. 8). The most important of these bridges was what the Catalans later called La Pont de la Carnisseria d'Avall. In time these bridges provided radial routeways from the core, allowing ingress and egress to all parts of the island via the eight gates in the walls. These gates were massive, tunnel-like structures made, like the walls, from locally quarried mares (sandstone) blocks. Within the line of the walls a lane ran right around enabling the rapid muster of soldiers to any part in case of attack.

In addition to these radial routes – the principal streets – was a complex network of secondary streets cutting across the main ones with more of a zigzag shape. The third type was a labyrinth of narrow lanes and alleys, usually dead ends, leading to private houses, streets probably similar to those found in the suqs of many Middle Eastern cities today. The streetscape would have been very varied, even chaotic to the untrained eye, and certainly quite unlike the regularity of the grid plan of streets that the Catalans were to introduce elsewhere on the island at a later date (García-Delgado Segués & Garau Alemany, 2000: 133). This built-up zone was bounded by the Riera to the west and the present-day locations of the Plaça Major to the north, the grounds of the convents of St Francis and St Clare and the sea, an area that was to form the core of the later, medieval/Christian parish of Sta Eulalia. It commanded a second high point made up of Senta Eulalia (not the present day church, leading scholars into much confusion!), a flight of steps known as Adderrachy, and what was probably a place of execution (beheading), Aqaba al Madbuh (Riera Frau, 1993: 92).

One feature that helped shape the morphology of the city, as it did in all Arab cities, was the water supply system, a feature important in urban areas because of its almost recreational and theatrical function. The Riera itself did not perform this role – it was more likely used as a means of waste disposal. Instead, long distance sèquias brought fresh and clean water into the city from qanats about seven kilometres away in the countryside. There were at least three such sèquias, the most important of which were those from Font de Canet (Arabic –qanat) and the font of the Emir (Ayn al-amir) (Riera Frau, 1993: 305–312). These entered the city from the north, branching out to serve all parts and to feed large aljubs excavated underground. Water was then distributed from a series of more than fifty, what in medieval England would have been called, *pants* or water spouts. (Eighteenth century versions of these can be seen in Inca – see Fig. 6. 8.) The dendritic lines of the water supply network are said to have influenced the layout of the Medina's streets. One of the best-known examples is the present day Carrer St Miguel, originally named by the Catalans Carrer de la Sèquia.

In functional terms, certain land uses helped determine the character of the city. The thirteen or fourteen mosques, for example, provided not only a religious focus to city life with regular calls to prayer, but their distinctive architecture, with its large rectangular shape surrounding a large open courtyard, had an important social role as a meeting place too. With the exception of two on the right bank of the Riera, all the mosques were in the 'high town' on the left. Similarly the four bath houses were also located there. The five cemeteries could be found on the city's periphery. It is likely that the area enclosed by the Arab/Berber walls was not fully built up. There would have been substantial areas given over to orchards, gardens, cemeteries and industrial activities (Barceló Crespí and Rosselló Bordoy, 2006: 31–43) (see Fig. 6. 9). Commercial and workshop activities are often referred to by name in the documents in great number, but it is quite difficult to trace their locations on the ground. We hear of silversmiths, metal workers, bakers, millers, potters, charcoal sellers and weavers of all kinds. It is likely that, as in the later medieval city, the residence of the craftsman and his family was coincident with his shop or trade outlet. The principal market or suq was also situated in the high town, near today's Plaça del Mercat that still carries its name, with a second market on the city's north-eastern edge, near *Bab al-Balad*. The city also had a distinctive Jewish quarter, probably from Roman days, to the north of the administrative district.

Until more archaeological research is carried out, it is difficult to speak with much certainty about housing in the Arab/Berber city. It seems sensible to assume that they had the familiar layout of Andalusian houses, fronting straight onto the street, with an interior patio and living, cooking and sleeping quarters beyond (García-Delgado Segués, 1998: 91–98). The Repartiment does give us some idea of numbers, divided on the one hand by those in the King's part of the city in 1229 and those in the barons' part, and on the other between occupied and uninhabited houses. In all there were about 3500 houses in the medina with 20–25% being unoccupied at the time of the conquest. It is not really possible to calculate the population of the medina from these data at this time, but García-Delgado, using a coefficient of 800–1000 persons to each of the city's ovens, computes a figure of 40–50,000 ((García-Delgado Segués, 2000: 106). This is surely an exaggeration, as most authorities believe the population of the whole island at the time of the conquest to have been about only 20,000.

By the end of the Islamic period, then, three fairly distinct zones could be identified in the townscape: first, the area of the citadel and the centre of administration containing the home of the military commander and ruler and the residences of the ruling elite, built as a symbol of power as well as of defence. This was surrounded by its own set of walls, ditches and gardens. Secondly, the *medina* itself, which included the major centres of Islamic culture – the markets (the suq), workshops, merchants' houses, mosques and baths. From here radiated the roads linking the city to its hinterland. Much of this area's social and economic activity was fashioned by the sèquias and the geography of water supply. Thirdly, on the periphery of the built-up area within the walls was a zone of *horts i jardins*, another major market, and the burial grounds.

Was it a planned city? Perhaps only in the sense that all urban areas show a certain deliberation – identifying the original site, putting up defences and centres of political and religious control and patterning the ground with social and economic land uses that reflected the culture of the inhabitants. Various early medieval Catalan sources confirm the internal organisational geography of the medina, suggesting a sophisticated administrative structure (Barceló Crespí and Rosselló Bordoy, 2006: 45–53). After 1229 a new culture would take over and begin to re-shape the city – and the whole of the island of Mallorca.

Medieval Mallorca, 1229–1519

La isla de Mallorca, en la centro de la cuenca del Mediterráneo occidental y casi equidistante de las costas del la península Ibérica, Itálica, o del Norte de África, ha sido testimonio del paso de navegantes que, desde la Antigüedad, cruzaron el Mare Nostrum con las repercusiones que este hecho pudo aportar al desarrollo de la Isla tanto a nivel económico como socio-cultural. (Barceló Crespí and Rosselló Bordoy, 2006: 15)
(The island of Mallorca, in the centre of the western Mediterranean basin and almost equidistant from the coasts of the Iberian peninsula, Italy and North Africa, has witnessed the passage of mariners since ancient times, crossing the Mare Nostrum that led to the economic and socio-cultural development of the island.) [Author's translation]

Majorca, which remains perhaps the most neglected of all the medieval Spanish kingdoms… (Abulafia, 1994: 3)

By the end of the twelfth century the Aragonese/Catalan kingdom had become a powerful unit in the western Mediterranean and in the north of the Iberian Peninsula. Under its youthful new king Jaume I (James I) who had been crowned in 1213 at the age of nineteen, it had long held territorial ambitions elsewhere. Whilst always seen as a hero to Catalans and Mallorcans, Jaume was in reality 'a serial adulterer, vindictive to his children, often cruel to his enemies and not over endowed with humility' (Smith and Buffery, 2003: 9) but he was convinced of the religious case for his actions, like many medieval kings. His Christian kingdom was intent upon rolling back the Muslim frontier wherever it could. This was partly for reasons of religion and partly for sound economic reasons (Fernandez-Armesto, 1987: 11–42). On the mainland the rich lowlands of Valencia and the valley of the Ebro were looked on with some envy. Out at sea, to the south-east of Barcelona, the Balearic Islands were seen as a Muslim stronghold that harassed and hampered Catalan trading, although the differences between 'trading' and 'piracy' were slight at this time. If they could be brought under Catalan control they could act as an important stepping stone for further eastward expansion to Sardinia, Sicily and the Italian peninsula and, importantly, continue the trade with North Africa. In addition to getting their hands on the trading benefits of Mallorca, many of Jaume's supporters expected to receive other commercial benefits, notably those derived from land and farming. Indeed, the conquest of Mallorca has been seen as 'a crucible of colonial experiment, in which the problems of adjusting the balance of indigenous and incoming populations, native and new elements of the economy were tried … A pattern was established which remained influential throughout the expansion of the Crown of Aragon' (Fernandez-Armesto, 1987: 18), to some even a model for the *conquistadores* of the Americas after 1492.

One of the major sources for the conquest of Mallorca by Jaume I is the *Llibre dels Fets* ('The Book of Deeds') written at that time as a series of stories, a kind of legitimisation of the invasion. It flatters the king into believing that it would bring 'more glory than if you were to conquer three kingdoms on land' (Smith and Buffery, 2003: 70). Once a plan of conquest had been agreed a fleet was assembled of 25 full ships, 18 *tarides*, 12 galleys, 100 *buzars* (a small single mast sailing ship) and many *galliots* (small galleys of 20 oars) drawn from many parts of the north-west corner of the Mediterranean. The invading army it carried consisted of about 680 knights and six to eight thousand infantry (Smith and Buffery, 2003: 80).

By the time of the conquest the Muslims of Mallorca, nominally under Almohad rule since 1203, were virtually independent under their *wali*, Abu Yahya. They numbered about 15 to 20 thousand, slowly built up from their original invasion of about 100 years before. As we have seen, the island was divided into fifteen administrative areas and was farmed on the clan–tribe basis via a series of alquerias and rafals. The Muslim settlers had especially favoured the mountain areas to the north and west with their better water supply. Their only city was the Medina Mayurqa. In the period since the beginning of the tenth century it is unlikely that many of the original indigenous population would have survived, so that the total population of the island was probably less than 20,000, giving a low density of about five per square kilometre.

After the conquest, two challenges immediately faced the Catalans and Aragonese: what to do with the defeated Muslims and how to convert the island from an Arabic system of landholding and settlement to a more feudal one. The conquest was to have profound and long-lasting effects on the Mallorcan landscape, mostly derived from a version of the process of feudalisation. The two cultures were separated by much more than religion.

Earlier historians believed the Muslims were expelled, enslaved, or at best converted. More recently it has become clear that many remained, especially to farm the land and to practise their crafts within the city: blacksmiths, shoemakers, bakers, etc. (Lourie, 1976: 633; Bernat y Roca, 1997). The earlier claim that as many as 16,000 remained behind is probably an exaggeration, but enslaved Saracens were needed to cultivate the land, especially in the early post-conquest years because, as we shall see, colonising the island with Christians proved more difficult than expected. A proportion of the Muslim population, perhaps as slaves, perhaps converted, formed an important economic function in this period of transition (Mas and Soto, 2004: 35–9). Within a century there were as many as 100

African Muslim merchants alone resident in Mallorca (Hillgarth, 1976: 361–2). At the time of the conquest Mallorca consisted of some 1600 Arab/Berber farms and settlements outside the Medina, mostly alquerias and rafals, giving a total cultivated area of about 115,000 ha, about one third of the island. An important question for the landscape historian is: how much of the imprint of the Muslim era survived into the Christian landscape after 1229? In the early years there was probably very little change, despite the division of the island by the *repartiment* (Bernat y Roca, 1997: 29).

However successful the military conquest, the major task facing Jaume and his supporters was the colonisation of the island. At one level this entailed the planning of divisions and boundaries, some of which would be inherited from Muslim estates – largely based on access to water supplies – but many would be new elements in the landscape. In order to understand the consequences of this and the new patterns and system of landholding, it is first necessary to digress slightly into the arcane world of the Mallorcan version of feudalism, the effects of which were to last rather longer in the island (and Spain) than in more northerly European countries. In northern Europe, the Frankish system, as it was known, saw land held directly from the king or a tenant-in-chief in return for knightly or other military service. In some cases, often in relation to urban governance, a fee farm could be paid instead, and in time many chivalric payments could be paid in kind or cash. In most of France and Britain it led to a fairly clear hierarchical system of land tenure. But consider the case of Mallorca: originally under Roman law up to the sixth century, then Visigothic to the tenth century, and finally Arabic or Sharia law until 1229, after which it came under an evolved system of Visigothic/ Frankish land tenure. Added to this was a new form of inheritance that eventually was to affect the way in which land was handed down and subdivided amongst heirs, something that would affect the actual patterning of the ground. From 1229 Mallorca can be seen as 'frontier' land, literally out-front of the Catalan/Aragonese sytems of landholding and law. In order for such a society to function – to conquer and convert – much higher degrees of freedom had to be granted to the new settlers than would have been the case on the mainland.

King Jaume took about 50% of the island for his share and then divided the remaining half amongst his principal supporters, both nobles and churchmen, their reward for providing the army and navy and largely paying for the conquest (Morro Veny, 2003) (see Fig. 7. 1). Of the cultivated farmland Jaume retained 33,739 ha, 66,535 ha were given over to seigniorial holdings and 13,679 ha were held by urban interests from the mainland (Morro Veny, 2003: 18). There were to be three types of tenure: freehold or allodial *à la Franks*, emphyteutic leases, and thirdly the more familiar knights-in-arms service. The first was acquired directly from the king and based on the fixed assets of land and property usually held by a senior tenant *in eternum* ('for ever'). The second were long leases, again usually 'for ever', and were paid for in money and a share of the crops or produce – *terre*

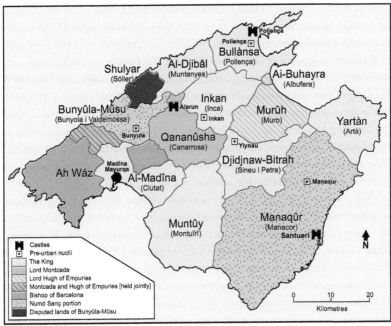

Figure 7.1 Map of *repartiment*, the division of Mallorca following the conquest by Jaume I in 1229.

et fructibus. There would usually be a 50% 'title' between the lord and the farmer, often accompanied by a tithe (*delme*) of one tenth and a tax (*tasca*) of one eleventh payable directly to the king. These leases were drawn up primarily for land from the subsequently divided alquerias and rafals taken from the Muslims. Hillgarth believes that most of the land grants in the King's moiety were for medium sized properties; less than one in twenty were for properties larger than 250 ha. Freehold agreements were at least as common as emphytutic leases (Hillgarth, 1991: 5). By the end of the thirteenth century the names of the holdings had largely been given Catalan forms. For the third type Mallorca was divided into *cavalleries*, an area of land varying in size but with 130 cavalleries providing approximately one armed knight. This measurement was used because the Crown had decreed that the island should provide 100 armed knights for its defence. So, 130 x 100 = 13,000 cavalleries. In fact, the island had 13,446 cavalleries. Of the 100 knights just over half would be drawn from the King's lands. As an example of this process, the Infant Peter of Portugal, uncle of Jaume I and Lord of Mallorca 1231–44 and 1254–56, received 103 agricultural estates, about 6100 ha, directly from the King as a tenant-in-chief, but it was never his intention to farm the land himself – '*aquests tenidors tampoc no treballaren la terra*' (Morro Veny, 2003: 25). In another example, in 1233 Nuño Sanc (one of the major conquistadors) conceded to Pelai Nunis, the cavalleria of Manacor, in return for one knight's service but Nunis, instead of paying in service, paid the money equivalent of 50 *quarteres* of wheat and 100 of barley, said to be equal to 50 *llibres* (pounds). The Count of Empuries was given 849 cavalleries equal to seven armed knights. Thus, a cavalleria could be derived from service or rent. In a third example, the town of Lleida in Catalunya held 198 cavalleries in Mallorca, consisting of 16 alquerias, in total 104 *jovades*, nearly 1200 ha. Even minor individuals were granted specific Muslim estates such as Pere de Sant Melio, the King's scribe, who was given the rafal of Toffay Aben Yusef (Vinas, 2004: 103). In the absence of the great barons who resided elsewhere, these cavallerias were the building blocks of the new Mallorcan rural system. The actual number of armed knights resident on the island remained small, since they usually translated their obligations into monetary payments or their equivalent, and the economic power rested with the 'citizenry' resident in Ciutat.

The Jews were important to the conqueror because of their trading function, acting for many years as intermediaries between the Christians and the Muslims, and were granted land and other privileges in recognition of this, in a charter dated 11 July 1231. In all they were allocated almost 1000 ha in the districts of Inca, Sineu, Petra and Montuïri, to include all land and buildings, meadows and pasture, water, woods and vines, together with all the output and revenue from the 18 alquerias and rafals granted them. In addition the Jews were also given a ten year exemption from contributing to military service and its costs (Vinas, 2004:254). Like the Christian lords, it is unlikely that they farmed these lands themselves, but sublet to new settlers. However, their true identity lasted only until 1391; after that date they were hidden as *conversos* created by forced baptism.

The major problem, however, was that there was no substantial body of peasants at the bottom of this hierarchy to actually provide the labour force for farming. Hence the need to retain some Muslims, if only in an interim period until Christian colonisers could be recruited. The fate of those Muslims who remained in the countryside is an unresolved question. The *Repartiment* only covered the Crown's holdings, about a third to one half of the island, but it would appear that Muslim farmers remained at least in the north, around Pollença (Abulafia, 1994: 58). The majority, however, were taken as slaves and forced to work for their new masters. In addition, taxes from Muslim farmers were an early important source of tax revenue (Lourie, 1976: 627). As Glick points out, however, most of the early settlers (and possibly some Muslims too) were in effect given 'free men' status as part of the drive to encourage immigration. This was especially true for the King's lands, where about 60% of the land grants were relatively small – between 25 and 100 ha – encouraging small settlers without the usual feudal controls (Hillgarth, 1991: 5). Gradually, however, the more familiar seigniorial patterns took over throughout the island as the century progressed (Glick, 2000: 34).

The significance of this detail about the distribution of land and its management is that it was to affect the distribution of cultivated land and the associated farms and large *possessions* (landed estates) for many centuries, with many of the larger holdings not broken up until the reforms of the late eighteenth and nineteenth centuries. As we shall see in later chapters, this was also to affect the rate and direction of shifts in the geographical margins of land colonisation as more and more land was brought into economic use.

Our major source for these changes are the *Llibres del Repartiment* which describe, as their name suggests, the division of the island amongst the new settlers, a series of land registers which 'document a cross-cultural transfer of landscape', according to Glick, 'a source for recreating the lost landscape … and for evaluating the ensuing modal change in the ordering of the countryside' (Glick, 1995: 128–30). In the case of Mallorca they survive in Arabic, Latin and Catalan, but regrettably have limited geographical coverage. Following Soto, in his analysis Glick goes as far as to say, 'the Christians changed the landscape immediately' (Glick, 1995: 133 and Soto i Company, 1990), into one with an emphasis on grain cultivation, with more of the water management systems given over to mills rather than irrigation. The farmed land in the early thirteenth century was largely located in the better-watered fringes of the highlands and in the south, with much less in the coastlands to the east or north or in the centre. Water for irrigation was a major determinant, and as in so many aspects of economic life, the 'conquistadors' initially had to work with what they inherited from their Muslim predecessors. However, we showed earlier that under the Muslims agriculture was dominated by pastoral farming over much of the island, with cultivation concentrated in the north and west and around the Medina. With the new feudalism and the Catalans/Aragonese came an important change: the growing of wheat on a large scale as the major bread grain, and the use of wheat as a source of taxation and rental payments. The relatively small irrigated gardens (*hortes*) of the Arabs and Berbers around the major settlements gave way to extensive fields of wheat and other grains, much of it on the central plains and in the coastal lowlands, in some areas grown between tree crops such as mulberries and figs the beginnings of a landscape feature that persists to this day. This usually involved the clearance of large areas of garriga, altering the 'natural' landscape for evermore, although this was a slow process continuing well into the early nineteenth century. Clearly, this was the beginning of a lengthy process of landscape transformation based on differing cultural and economic values translated into patterns of land-use.

Water Mills

If the spread of grain farming through the thirteenth and fourteenth centuries brought about a significant shift in the agrarian landscape of Mallorca, then the location and distribution of water mills was complementary. Although the new settlers built additional water mills to increase the output of flour, many sites were inherited from their Muslim predecessors. Like tenure, landholding and crops, mills too became feudalised. From the middle of the thirteenth century more and more came under the control of the seigniorial lords, who saw them as a source of income. For example, the mills in the Alaró area fed by the *qanats* of Les Artigues and La Bastida that we described in the previous chapter originally came under the control of the Count of Béarn, but were granted by him to the Knights of St John, who further sublet them to tenants later in the thirteenth century (Riera Frau and Soberals Sagreras, 1992). Initially, the first settlers followed the existing Muslim pattern, but more and more mills were built along the more reliable watercourses after qanats were lengthened to help guarantee a better supply of water – '*fonts novas*' as opposed to '*fonts vells*' (Carbonero Gamundi, 1992: 45). Another example can be found at La Granja (Esporlas) where the qanat de l'Ullal was driven 230 m in the 1240s and its sèquia supported 32 mills – '*la major concentració de molins de tota Mallorca*' (Gorrias i Duran and Terradas i Jofre, 2005: 314). Many factors influenced mill location, including the physical capacity of the sèquias and adequate breaks of slope in the long profile so that leats could be constructed to supply the *molins de cup*. But proximity to markets was important too, so that, for example, of the 62 mills mentioned in the *Repartiment*, 47 were to be found on the sèquias flowing into Ciutat especially on the Sèquia de Canet. Another major concentration on the east side of the island was along the axis between Canyamel and Artá (Carbonero Gamundi, 1992: 40). Among surviving contemporary structures are the linked water mills of the Galatzó *possession* in Calvià (see Fig. 6. 5).

Ordinacions of 1300

Despite Glick's supposed 'immediate' transformation of the agricultural landscape, in fact the process must have been somewhat slower, largely because of the difficulties in attracting new settlers. Early medieval demographic statistics for Mallorca are notoriously weak, and various authors' estimates vary widely, but if the Muslim population numbered less than 30,000 at the beginning of the thirteenth century, by its end the 'new' population had probably risen by only another 10,000. By the time Jaume I was succeeded by his second legitimate son in 1298, it was clear that a more adventurous policy for colonisation was required if Mallorca's land resources were to be properly exploited. [1]

Jaume II drew up a most remarkable series of documents that were dramatically to alter the island's landscape. These were the Ordinances of 1300, a series of plans to colonise the island based on a number of planned nucleations, in effect, planted towns (Fig. 7. 2). The model for the King's plans was probably the *bastide* town of south-west France and Wales built by the English Plantagenets, and the settlements founded by his father in Catalunya, both part of a Europe-wide movement to build new towns in the Middle Ages (Beresford, 1967). Although originally assumed to be based on defensive requirements, the modern interpretation is that they were primarily economically motivated, to stimulate development in new areas. Indeed, the architect/planner Alomar saw them, in the language of French regional economics of the 1960s, as *pôles de croissance*, growth poles (Alomar Esteve, 1976).

Their locations were to be almost all in the *part forana*, the area that had benefited least from new Christian settlement and the population growth in the latter

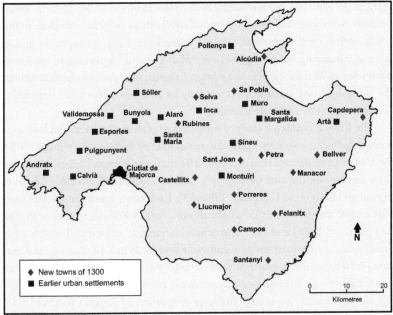

Figure 7.2 Map of the new 'towns' established under the ordinances of Jaume II in 1300, designed to act as economic and settlement centres to attract new settlers.

part of the thirteenth century, which had been concentrated in the mountains, the *raiguer* and to a lesser extent, the *mitjana* where the expansion of cereal growing and water mills could take place, thanks to a better water supply situation. The lands in the centre of the island and in the east were given over to *ramaderia extensiva* at this time, the grazing lands of the former Berber settlers. Much of it was covered in *garriga*. Overall, it might be considered as 'frontier' land, literally out front of the new Catalan and Provençal settlers. These dry lands also formed a large part of the King's property, with as much as 72% located in *la marina*, and only 14% in the mountains and the *mitjana* respectively. There was clearly an incentive for the Crown to increase exploitation of its own land resources. As we have seen, the estates of the powerful barons and churchmen had already taken the 'best' lands. Thus, it was likely that any new colonisation policy would have to be aimed at smaller farmers and designed to offer them workable parcels of land, a water supply, defensive security, markets and churches, as well as housing, and to give to merchants and tradesmen certain rights in an urban setting. All this pointed towards a need for small *towns* or at least substantial nucleated villages.

Earlier research suggested that eleven towns were built – Algaida (Castellitx), Llucmajor, Porreres, Felanitx, Campos, Santanyí, Sant Joan, Sa Pobla, Manacor, Binisasalem (Robines) and Selva – to which were added Petra (Andreu Galmés, 2000) and Sineu with certainty, although more recent work believes another three were begun, but for which there is no record. Many were located at the site of the larger alquerias, often retaining their Arabic names in a Catalan form. Some definitely were given a defensive purpose, largely those to be located close to the coasts open to continuing corsair and pirate attacks such as Capdepera, the hilltop town commanding the passage between Mallorca and Menorca, and the north coast town of Alcúdia, close to the Roman *Pollentia*. One curious characteristic of the location of these settlements is that they are all about twelve kilometres apart, surely the result of deliberate planning.

The King appointed a team of surveyors to plan the location and layout of these new towns – we even know the names of their leaders – Pere Estruc (who died in 1304 and was replaced by Bernard Beltran) and Ramon Desbull. They also drew up plans for the vital water supply for each town in a *Codex Liber Aquarum Forentium* (Carbonero Gamundi, 1992: 47). Each town was to house 100 families on 500 quarterades (355 ha) of cultivable land, 1000 quarterades of garriga and pasture (710 ha) and 12. 5 quarterades of horta, all to be distributed to each family as follows: 5 quarterades of cultivable land (3. 55 ha), 10 of garriga and pasture (7. 10 ha) and 0. 5 quarterades of horta (0. 36 ha). Paid labourers were to be allocated between one and two quarterades of cultivated land (see note on p. xiv for Mallorcan measurements). The large proportion of garriga ('frontier land') seems to represent an expectation that the new immigrant settlers would expand the margins of grain cultivation, necessary for the ever-present challenge of feeding the island's population and, eventually, helping to fill the King's coffers. However, in order to attract people to these sites they were given a three year moratorium on payment of dues and taxes, with monetary and cash payments being possible later in lieu of military service – *monititizació de la renda* or money rents (Andreu Galmés, 2002).

The driving force behind this burst of town building was social and economic and had more to do with repopulating sparsely populated areas '*que es una repoblació de territoris erms o abandonament*' (Andreu Galmés, 2000) rather than 'bricks and mortar', although some have seen the influence of St Thomas Aquinas's *de regimenine principium* at work: '*The city is the perfect community … and building cities is the duty of kings … The most powerful nations and most illustrious kings acquired no better glory than comes of founding new cities.*' Perhaps it is better to see them simply from a multifunctional point of view. The nucleations were to be based on the parishes established after 1229, often on old alquerias, but some developed around churches, others in new spaces between existing, much older settlements. An example of the latter is Llucmajor located between the Arab centre Son Granada and the Talayotic settlement of Capacorb.

The internal structure of the towns was to be that of the grid plan, following a long history of this type of formal layout, derived from the same Roman surveying techniques employed in centuration (see Chapter 5). The urban morphology consisted of a network of streets, blocks (*illetes*) and buildings with each house occupying a *quarter de terra* (1775 m²). Each illeta was made up of four *cairons* (lit. 'tiles'). Within these rectangular spaces room had to be found for churches, markets and squares. The latter were usually to be found at the centre of a set of illetes, sometimes in front of the church as at Petra, or fronted by inns as at Felanitx (Fig. 7. 3). This, then, was the *planned* morphology but, as with such towns in other parts of Western Europe, it often took centuries for the building units to become occupied. Figure 7. 4 gives some idea of the original plan and its expansion by the late eighteenth century for Llucmajor. For many of these towns it was not until the beginnings of more modern urban functions associated with an urban milieu in the eighteenth century that they began to resemble what we see today. In any case, the Mallorca of the first half of the fourteenth century saw little movement of population to these towns. And then the Black Death struck, a demographic disaster that was to have profound effects on the economy and the landscape.

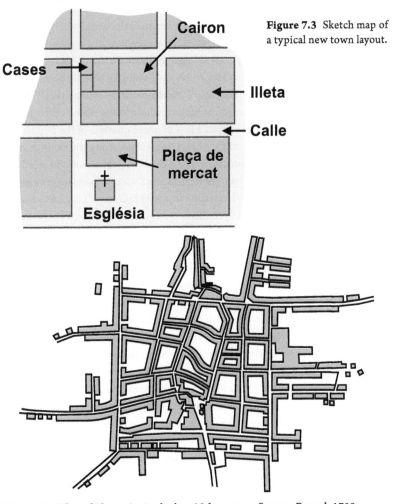

Figure 7.3 Sketch map of a typical new town layout.

Figure 7.4 Plan of Llucmajor in the late 18th century. Source: Berard, 1789.

The Black Death and its effects

Probably brought westwards by Italian merchants from its origins in the Black Sea area, and particularly the port of Kaffa (Feodosiya), it reached Marseilles in 1347 and by December of that year it had entered Mallorca – unsurprisingly early, given the island's location at the heart of west Mediterranean shipping routes. A certain Guillem Brassa was said to be its first victim on 'Spanish' soil; how many others were to follow his fate is the subject of much debate among scholars of the plague. Certainly the full effect was felt in the second half of March and in April 1348, but by the end of May it had begun to subside; but a series of plagues followed, certainly well into the fifteenth century.

It is important to have a fairly accurate estimate of the loss of population due to plague because of the varying economic effects, and hence its subsequent impact on the landscape. Any calculations require 'before and after' data, and although what was known as the Black Death in Europe begins in the late 1340s, it continued for varying periods after that date. Hillgarth, for example, initially gives Mallorca's population as being reduced by 60% (Hillgarth, 1976: 363) but his calculations based on 'morabatí' data (see p. xv) for 1329, 1343 and 1350 show a decline in the latter seven year period of about 10,000 or 20% of population (Hillgarth, 1976: 12). Bisson quotes the Crown of Aragon lands as a whole losing 60% of their people (Bisson, 1986: 163) and Cohn reports a loss of 'three quarters of the population, all in less than a year' (Cohn, 2003: 111). Benedictow, using the same data set as Hillgarth (1343–50), estimates a fall of only 4% in Palma, but where 37% of households liable to the morabatí tax were located. Countryside losses amount to 23%, to give a very low overall mortality loss for Mallorca of only 16%, a figure remarkably similar to that used by Santamaria a generation before (Benedictow, 2004: 1). He also notes that the number of Jewish households liable for morabatí actually rose in this period from 333 to 516, perhaps supporting the urban/rural contrast noted above. In general, then, what Benedictow seems to suggest is that it was the increase in Mallorca's population numbers just prior to the Black Death that was wiped out, leaving totals at the end of the 1340s not very different from those at the beginning of the decade. If Mallorca escaped so lightly from the Black Death, which carried off more than fifty million people in Europe, was it because it was a highly urbanised island with over 50% of its population in Ciutat? Hypotheses for the causes of the Black Death are numerous, and recently bacteriologists and archaeologists have sought to question the longer-standing rat/flea causation. As Benedictow points out, the ratio of human beings to rats and their fleas will tend to be lower in urban environments than in rural areas; that is, in the countryside there will be more people to share the rat fleas between them. When we consider the level of mortality from the Black Death in terms of its eventual economic impact, the figure of 16–20% for Mallorca seems somewhat low.

In the decades following the 1340s and '50s there was a marked shift in the agricultural economy of Mallorca from the grain farming that had been emphasised, towards sheep farming, one reason for this being that pasture needed considerably less labour than grain farming. However, economic changes are rarely explained by one fact alone, and the growing demand for woollen cloth throughout Europe played a significant part too. Gradually some grain production gave way to extensive sheep ranching so that more land was brought into agricultural use, but less intensively. With most land remaining in the hands of the absentee seigniorial class, new four to five year leases were drawn up to support this new agricultural activity, particularly in the dry area of the island, especially around Artà, Inca, Pollença, Petra and Manacor. However,

> En una zona con suelos y características climáticas poco adecuados para cereales, el cultivo de los mismos no era garxantía de obtener et grano suficiente para las necesidades de la economía campesina.
>
> (In a region with soil and climatic characteristics little suited to cereals their cultivation was unlikely to meet the needs of the rural economy.) [Deyá Bauzá, 1997: 78]

In turn this new source of wool stimulated a nascent textile industry, the other condition for success being the tension between the guild structure (essentially based in Palma) and an emerging merchant class who sought to control quality from the raw material stage through to the finished product. Mallorcan cloth was for export and the trade was an important source of revenue for the purchase of wheat, of which the island was constantly short. This industry also created new employment opportunities, initially in Ciutat, but later in the relatively new 'new towns' of 1300. Except in the towns, its landscape impact was relatively small. Despite constant visitations of the plague throughout the fifteenth century, the island's population grew from 23,510 in 1457 to 31,525 thirty years later (Deyá Bauzá, 1997: 47).

Woodland and trees in the medieval landscape

Naturally, conflicts arose between the different land uses: pasture for sheep, grain cultivation, and the tree-based economy of the uplands and elsewhere. In landscape terms this meant there was a moving frontier between them partly determined by prices (supply and demand) and changing climate and weather, and controlled by soil and water conditions. This process of fluctuation was most noticeable on the lower slopes of the mountains and where farmland met woodland in the accidented relief of the central lowlands. Woodland played an important part in the life of medieval Mallorca. It was a source of raw material for buildings, for shipbuilding, for machinery and furniture and for charcoal-making as well as a place of pasture and hunting.

In order to appreciate this role, we must first try to agree on what was and what was not 'woodland' and who had access to it at this time. We showed in two earlier chapters that dealt with vegetation (Chapter 3) and prehistory (Chapter 4) that the idea that Mallorca was once covered in dense and continuous forest, cleared by early societies, was almost certainly erroneous. There is a considerable debate about the nature of medieval woodland, and earlier views about 'forest' have now given way to accepting that there was a variety of types of woodland, some the result of environmental changes and some of human activity that itself varied over time. By the thirteenth century and up to at least the mid-eighteenth century Mallorca was probably characterised by two main types: the mountain woodlands dominated by pine and holm-oak and a quasi-savannah landscape elsewhere in which tree cover varied but was part of the garriga, and in other places interdigitated with cereal crops and pasture. In the case of the latter, it is likely that in the dry areas of the island where medieval sheep ranching was established in the fourteenth and fifteenth centuries, a much more 'open' kind of savannah would have obtained, since grazing would have restricted regrowth in land that was cleared. Where surface water was more readily available and arable was more dominant, then trees would have been more abundant. In the case of both types of tree cover, it is almost certainly true that there were fewer trees in the medieval landscape than there are today.

Away from the Tramuntana in the fourteenth to sixteenth centuries the dominant vegetation would still have been the garriga, that low-growing combination of Mediterranean plants (lentiscus, cistus, etc.) interspersed with lowland trees. It was this that was to be 'cleared' progressively over the next three centuries to produce much of the more open landscape that was to receive the new tree-based economy of the eighteenth century – almonds, carobs, figs, vines and olives, although the later were essentially a upland crop. 'Cleared' does not mean wholesale eradication of trees, but rather their selective removal to produce what Rackham calls the 'cultural' savannah (Grove and Rackham, 2001: 194). This is the kind of part-wooded landscape that would have eventually been created from the 1300 ordinances, where large sections of garriga were allocated to the new settlers who, rather like settlers in North America at a later date, were major agents of landscape change through clearance and management.

By the time of the Christian conquest much woodland remained in the mountains in the north and west, despite previous Arab/Berber activity in creating terraces for agriculture. Here, wood was a valuable resource and would have been very carefully managed and not removed wholesale, as an earlier generation of historians would have led us to believe. McNeill's assertion that the Spanish mountains were denuded by shipbuilding, wood-fuel, charcoal making and construction was certainly not true of Mallorca; it would have been a fine example of irrational economic behaviour if it were (McNeill, 1992: 37). Various edicts were issued from the Crown to regulate the cutting of timber because of its strategic importance (Gil et al., 2003: 118–9). Only where a more profitable land-use might have replaced woodland would it have been subject to continuous clearance. To most modern authorities, medieval woodland is seen as a kind of 'crop' to be cultivated, harvested and occasionally, when necessary, replanted. This 'crop' in medieval Mallorca was largely pine of various kinds and holm-oak.

Medieval Mallorca made its living by trade, trading with most parts of the western Mediterranean, especially with North Africa and even as far afield as Britain and the Low Countries. David Abulafia has described the island as a veritable emporium (Abulafia, 1994). It is not surprising, then, to find the island engaged in shipbuilding, an industry heavily dependent upon timber. The fourteenth century may have been the golden age of Mallorcan shipbuilding, but the ships built at this time – *galeras, galeotas, llenys, corses, gorabs* and *taridas* – were quite small and of simple construction (Gil *et al.*, 2003: 111). Local woodlands could supply certain ships' timbers, particularly from pines and *encinas* (holm-oaks), but by no means all, and from an early date Mallorca relied on imports. Once the longer Atlantic sea routes were opened up in the sixteenth century, then Mallorca was poorly positioned to benefit from much of this trade, and its 'inland sea' ships proved unsuitable for transatlantic voyages. Except for a short period, then, it is unlikely that the island's shipbuilding industry would have had a detrimental effect on its woodlands. Similarly, the Mallorcan woods were able to supply some timber for building, including fortification, and furniture, but much of it was unsuitable for the large public buildings of the period such as Palma's cathedral and the numerous medieval churches.

Charcoal emerges in the late medieval period as an important fuel, having a higher calorific value than raw wood and able to withstand higher transport charges from the uplands to Ciutat. However, charcoal is made from branches and small timber, not the massive trunks of the 'virgin forest'. The source of wood for charcoal was primarily from coppiced and pollarded trees, very much the 'crop' referred to earlier, and often harvested from the same places well into the twentieth century. McNeill's calculations may well be correct in stating that 'a modest village of 100 families needed reliable access to 200–400 hectares of tall forest', but less accurate in believing that 'charcoal use *destroyed* forest at 2–5 times the rate of direct wood burning' (McNeill, 1992: 136–7). Whatever the strength of the respective arguments, charcoal burning has left its mark on the Mallorcan landscape, although most of the *sitjas* – the round platforms upon which the charcoal furnaces were erected, which are discernable today – are of much more recent origin but distributed in similar areas of the Serra Tramuntana (Fig. 7. 5).

The pines of the mountains were also the source of pitch or tar, a quasi-industrial activity, but the amount of wood and the way it was used would hardly have contributed to the wholesale depletion of the pine woods. The 'distilleries' (*hornos de alquitrán*) can sometimes be found today as small-scale landscape elements. More common was the *forn de calc* for lime burning (Fig. 7. 6).

In summary, then, the wooded lands of Mallorca in the Middle Ages were a significant landscape feature with the greatest densities to be found in the mountains, as they are today. The trees in the landscape of the rest of the island were best seen as part of the garriga. From the evidence available there seems to have been little approaching the 'forests' or 'parks' familiar in more northerly European

Figure 7.5 Reconstruction of a *sitja*, a site for charcoal burning. These are found in many parts of the Tramuntana, often as large circular platforms. Such landscape features may date from the medieval period; they were used up to and including the mid-20th century.

Figure 7.6 A *forn de calc* used to produce quicklime.

landscapes where kings and lords reserved certain rights to woodland (Rackham, 2000). Similarly, there seems to have been a dearth of common land for wood to which peasants would have had access. The relative absence of woods in the lowlands of Mallorca meant that when economic pressure forced the agricultural frontier outwards to create more land for grain and grazing, it was the garriga that was cleared rather than the wooded lands at higher altitude. Trees were too precious a commodity that had to be carefully managed. By the end of the fifteenth century agriculture still occupied a relatively small proportion of the island's area. It was in the next, early modern period that these clearances accelerated and the rural landscape began to take on some of today's more familiar appearance.

Ciutat – medieval Palma

If agriculture and the exploitation of woodland were relatively limited economic activities in the Middle Ages, and if the urban system being established via the *ordinacions* was in its infancy, then the same cannot be said of Ciutat (Palma). Outside observers and visitors to the island would have known little of the central plain, the mountains or the coastlands, but they would have been aware of medieval Palma, one of the great nexuses of western Mediterranean trade. To all intents and purposes the city of Mallorca – Ciutat – *was* the island.

In the 200 years following the conquest the city was to develop the principal characteristics of its townscape: its cathedral, the new churches, the conversion of the Almudaina, the Llotja, the first bishop's palace, the major trading streets and their attendant merchants' houses and early industrial developments – eventually to be surrounded by great new protective walls, the third of a number of encirclements (Barceló Crespí and Rosselló Bordoy, 2006) (see Fig. 7. 8).

Following the conquest the medina, rather like the island as a whole, was divided into two parts with the king having one half and the other divided between

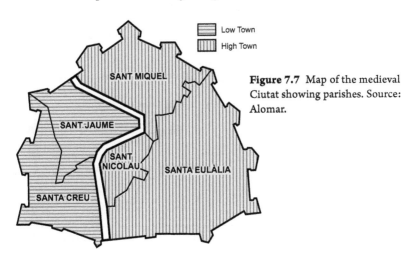

Figure 7.7 Map of the medieval Ciutat showing parishes. Source: Alomar.

the barons and bishops. The king's moiety coincided roughly with the 'high town', that is the area to the east of the Riera and containing the important signifiers of power – the Almudaina, the great mosque, the sea gate, etc., whereas the 'low town' was given to the six or seven *grandes magnates* (see Fig. 7. 8). What the Christian conquerors inherited was, of course, a city whose morphology was derived from Arab/Berber urban principles, and it was difficult to eradicate their influence overnight. The basic street plan within the twelfth century walls was one of narrow alleyways, often dead-ends, rarely more than four metres wide, with Arab style patio houses presenting blank façades to the street; even by the nineteenth century over 70% of the city's streets were less than four metres wide. The major public buildings of the Almudaina and the important mosques would have dominated the skyline, the port at the mouth of the Riera protected by the huge sea gate. Initially there was little opportunity for the Catalans to impose their own ideas of town building onto this form. The early task for much of the thirteenth century was to adapt or eradicate the Muslim features, the most obvious example being the decision of Jaume I to build the massive gothic cathedral to replace the major mosque of the Medina, perhaps the most dramatic symbol of

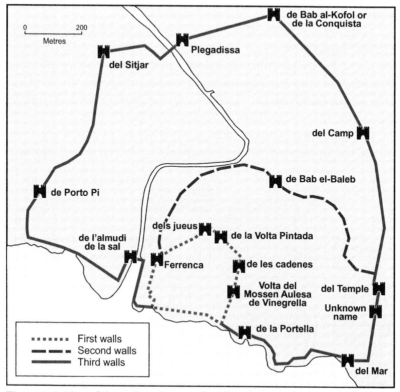

Figure 7.8 Map of the walls and gates of medieval Ciutat. Source: adapted from Barceló Crespí and Alomar.

MALLORCA
VISTA DE LA CATEDRAL Y CAPITANÍA GENERAL
Librería Escolar. P. de Cort 12.-Palma de Mallorca

Figure 7.9 La Seu – the cathedral of Mallorca, founded by Jaume I on the site of a former mosque. One of the largest medieval cathedrals in Europe, it has been added to continuously. Many of the features seen today date from the 17th century.

conquest and culture change visible to all who entered Ciutat's harbour to trade. Of course, the building we see today is the product of many subsequent centuries (Fig. 7. 9). The Medina's walls had been badly damaged during the siege of the city, and it would be some years before new defences were constructed. The basic water supply system built by the Berbers was easily adapted to serve new needs, although new 'rules and regulations' concerning distribution via sèquias from the Font de la Vila had to be drawn up (Ginard Bujosa, 1995: 31). Some public spaces were soon taken over for new activities, with the plaça in front of the mosque that stood where the Catalans built the parish church of Sta. Eulalia in 1248, for example, being converted to a marketplace.

Thus for about a century the new masters of Mallorca were only able to impose limited new 'land values' on an Arabic urban landscape; but gradually between about 1300 and 1500 a recognisable European medieval city emerged from its Roman and Arab origins. It is worth recalling that almost immediately following the conquest, merchants from many parts of the western Mediterranean 'set up shop' in Ciutat; the men of Marseilles who had aided Jaume I alone were granted 297 houses. They were followed by Pisans, Genoans and returning Jews. For most of these new groups, unfortunately, we do not have any contemporary maps to guide us as to where they settled. Similarly, there was very little new by way of town planning, much of the mosaic of inherited streets and blocks proving too persistent because of the economic value already embedded in them.

Like nearly all early medieval cities, there was a considerable amount of open space within the walls – space that was originally used for quasi-agricultural activities such as gardens and orchards, piggeries, animal pens and middens. Many of these spaces derived from the Arab's *'horts i jardins'*. Some, such as the convent of St Francis and the monasteries of St Elizabeth and St Clare, were taken over by the incoming nuns and monks (see Fig. 7. 7). Much of the land on the periphery of the city, but within the walls, had to be kept open to give ready access to the walls if under attack. As population numbers grew, some of these areas were built on. Two such areas were in the south-east and south-west corners of the city where two new *barrios* were built – Calatrava and Puig St. Pere respectively – which displayed elements of the more familiar rectilinear street pattern of the fourteenth century, something we observed in the 'new towns' of the 1300 ordinances. A major addition to the street plan was the development of four major radial roads that led out from the core. One, from Plaça de Cort to the Sta Margalida gate, linked the city to the Tramuntana. Another, from the same place to the St Anthony gate, contained important shops and workshops and led out to Es Pla and Manacor. A third ran eastwards via the Camp gate towards Llucmajor, and a fourth crossed the Riera, linking the city via St Cross parish and the Sta Catalina gate to the west of the island (Barceló Crespí, 1988: 71). Both the gates and the river were important morphological elements. The former controlled entry and exit, often a place of tension between Ciutat and the Part Forana. The latter, besides dividing the city into its upper and lower sections, had to be bridged in order to carry these new routes. The Riera was a far from stable element, with massive floods in 1403, 1444, 1618 and 1635 that caused considerable loss of life and much damage to the evolving city. It was not diverted until the seventeenth century. In the medieval period seven or eight bridges were built to link the low town to the west of the city with the high town to the east, improving intramural communications. As the medieval city spread away from the old core of the Medina it moved onto higher ground, and one new morphological feature that emerged to link the 'low town' to the 'high town' was the stairway, an efficient form of thoroughfare for a city that had little wheeled traffic. Examples include Corralasses and Midonera in Puig St Pere, which may even be Arabic in origin. In later medieval times a fish market stood on the hill that is now crowned by the Plaça Major, and it too has stairways leading to it, including Costa de Can Poderos and Costa d'en Sintes.

We saw earlier that the creation of parishes and the building of churches in the countryside and in the new planned settlements was part of the *Christianising* of the landscape (see Fig 7. 7). In the city the cathedral, the five new parish churches, and a network of secondary churches were built, many in the dense set of streets running off the Calle St Miguel in the 'high town'. Some of these churches, such as St Miguel itself and Sta Eulalia, replaced former mosques. Many others were associated with another new townscape element, the convent or monastery, forming freestanding *'illetes de camp à l'interior de la ciutat'* (small islands of countryside in the city) (Barceló Crespí, 1988: 67). Their origins lay in the mother houses in mainland Europe and were part of the spread of this kind of religious activity often devoted to charitable works in medieval urban areas. The Franciscans followed Jaume I almost immediately in 1232, although their magnificent church and its cloisters were not started until 1281. The Augustinians eventually took over Santa Margalida convent, originally founded by the Franciscans in 1238. Some orders constructed hospitals such as Sta Catalina for elderly merchants and sailors. Among the military orders the Templars, who had taken part in the conquest, built their castle on the site of the Arabic Almudaina de Gomera against the east wall. Between them the churches and the convents added new townscape elements to Ciutat that were significant in more than material terms. New additions were also developed in the sixteenth and seventeenth centuries by the Jesuits and other later orders. Many were to disappear following the drastic secular reforms of the nineteenth century, when many convents and monasteries were dismantled and their lands put to other uses. But today there are sufficient relict features of this long Christianising era to remind us of the cultural changes wrought by Jaume I and his successors, townscape features whose persistence reflects a symbolism beyond bricks and mortar. [2]

The Jewish community predated the Christians in the city by many centuries, possibly from Roman times and certainly from the fifth century AD. They too benefited from their assistance to Jaume I in the lands granted to them elsewhere on the island. In Ciutat they were allowed to develop centres in the city, known as a *call* (see Fig. 7. 10). The *call menor*, to the south of present day Plaça Weyler, was centred on Calle Sant Bartomeu. This was later mostly subordinated to the larger

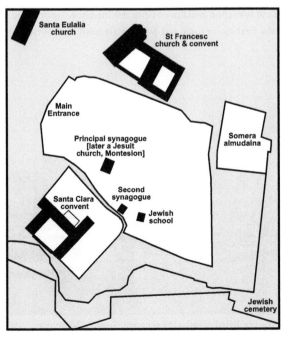

Figure 7.10 Map of the *call*, the Jewish quarter of medieval Palma. Source: adapted from Barceló Crespí and Rosselló Bordoy and Picornell Bauçà.

Santa Eulalia church

St Francesc church & convent

Main Entrance

Principal synagogue [later a Jesuit church, Montesion]

Somera almudaina

Second synagogue

Santa Clara convent

Jewish school

Jewish cemetery

call major focused on Carrers del Sol and de Montisión, between the Convent of Santa Clara, the Franciscan monastery and the Temple. This much larger area was to replace the more northerly section later taken over by the Dominicans (Barceló Crespí and Rossello Bordoy, 2006: 159–163). This *call* contained synagogues – the most important of which stood on the site of the present-day Montisión church – and large merchant houses, and their cemetery was located just outside the walls. Such a quarter (*aljama*) was probably not a ghetto initially but perhaps an *eruv*, within which Jews could practise their way of life in public as though they were at home.

Amongst its diverse economic activities Ciutat's primary business was, of course, trade, and although this was to fluctuate over the medieval period, mercantilism left a distinct mark on the city's landscape. Two organisations characterised this function: the College of Merchants founded in 1403 and the Consulate of the Sea founded in1326, each with its own iconic building – La Llotja (perhaps Mallorca's most beautiful building) built in 1426 (Fig. 7. 11) and the Consolat opened in 1669, both located at the heart of trading activity on the harbourside. A whole new suburb grew up just outside the medieval walls between the mouth of the Riera and Porto Pi – *el raval de mar* (Barceló Crespí, 2012). Two factors influenced the location of trading: proximity to the harbours and the availability of land for warehousing, storage and a whole variety of activities associated with shipbuilding, repairs and provisioning. Under the medieval rulers two principal ports operated in the Bay of Palma: Porto Pi to the west and the main harbour originally located at the mouth of the Riera. There was no complete sea wall at this time, and it is likely that land was soon reclaimed from the foreshore upon which ships could be beached, the first stone quay not being built until the fourteenth

Figure 7.11 The Llotja – the medieval merchant exchange. Source: Muntaner.

century, to the west of the mouth of the river (Barceló Crespí 2012: 17–35; Soler Gaya, 2004: 62). Shipbuilding was clearly an important activity in Mallorca, although on nothing like the scale of Barcelona or Genoa at this time, and as we have shown, local suitable timber was in any case in short supply. The wide variety of types of vessel constructed – nearly all small – meant that some could be built under cover. A large arsenal for this purpose was built at Sa Drassanes to the west of the Riera, a name preserved in the *plaça* of the same name.

Trade defined Ciutat, but as Abulafia has observed, 'What we are looking at is the transformation during the fourteenth century of a commercial entrepôt into a centre of production' (Abulafia, 1994: 106). Nearly two-thirds of the active population was in manufacturing by the late fifteenth century, with the city's industry dominated by textiles and associated activities as the following figures of numbers engaged in certain trades show:

Table 7. 1 Textile industry in Ciutat (Palma).

	Weavers	Cloth merchants	Tailors	Dyers
1478	104	267	52	20
1512	159	414	65	8
Source: Barceló Crespí and Rosselló Bordoy, 2006.				

Other activities developed, probably again from their Muslim forebears, in order to supply the city, which contained about 40% of Mallorca's population, and to a lesser extent the countryside, with goods and services. Initially workshops would have been widely spread through the city, but gradually areas of specialisation began to emerge. Some industries were located away from the city centre for environmental reasons so that Calatrava, for example, had a concentration of tanners and leatherworkers. We have shown the dependence of the island on grain imports, and much mercantile activity and milling would have been located in the city; windmills persisted in the cityscape well into the twentieth century. When the wool textile industry began to develop in the 1400s its primary location was in Ciutat, supported by a powerful guild structure. Spinners, weavers, dyers and those engaged in trading in raw wool and cloth operated from premises that combined the master's house with workshop or retail outlet or warehouse, each adding new elements to the urban morphology, often in very cramped conditions. The textile trades were concentrated in Sta Eulalia parish with over 40% of the 414 cloth merchants in 1512, well over half of the tailors and wool weavers and about a third of the linen weavers (Barceló Crespí, 1988: 115). Few of these remain today, but street names sometimes give a clue to the location of other trades, such as Calle Panes (bakers), Calle Pescateria (fishmongers), Calle Carnisseria (butchers), Argenters (silversmiths) and Videria (glassmakers).

These activities, together with the early immigration into Ciutat, began to bring about an increase in the city's population. Calculations have been derived from a number of taxation sources of the medieval and early modern eras, including *morabatí* and *talles*. The former counted heads of household with wealth of over 10 lliures (pounds). The latter tax exempted the poor, religious houses and the nobility, and was based on hearths. Generally a coefficient of 4 or 5 is used to convert these to actual population. This gives Palma 24,515 inhabitants in 1329, rising to 33,505 in 1573, with over half of them living in Sta Eulalia parish, with the more peripheral parishes having about 15% each. Over time, as the city grows the relative proportion in the central areas declines and the outer zones increase, probably the result simply of growth, although whether by internal migration from the countryside or internal rearrangement, we do not know.

Conclusion and prospect

In many respects the medieval and early modern eras are amongst the best known to us, thanks to the archival work of a number of distinguished historians. From their work it has been possible to paint a picture of Mallorca at this time that shows a landscape of a slowly expanding agriculture, easing the frontier back into the garriga, the hills and the mountains, a process that will accelerate from now on. The local feudal system largely eradicated much of the landscape it inherited from its Muslim predecessors, although place-name elements remain as a reminder of what went before. This notion of the 'frontier' was emphasised by Jaume II's policy of town building in the countryside, bringing a new feature to the appearance of Mallorca. Much of the coastal lands remained uninhabited or at best with low densities of population. It took a long time to overcome the ravages of the Black Death and other plagues as a depleted population took on a new economic value. Industry and trade grew rapidly in the fifteenth century, much of it focused on the primate city of Ciutat, which by the end of the century was a major trading port in the western Mediterranean. In the early modern period that followed, Mallorca was to lose some of its predominance in this field, and the new Spanish Empire would divert attention to another sea, the Atlantic. The old regime that began with the Christian conquest would persist for at least another two centuries until the Enlightenment began to have its effects on the economy and social organisation of the island. Any growth was slow indeed, and the landscape retained most of the features that had developed in the period we have just studied, despite the fact that more and more urban capital was invested in land rather than trade. It was to remain a place that continued to find it almost impossible to feed itself as it continued its age-old struggle with a harsh environment of poor soils and water shortages, not helped by a society held in the thrall of reactionary elements.

Early modern Mallorca, 1520–1820

But the greatest riches and wealth of this island, is from the olives; which yield an extraordinary quantity of oil. It is very remarkable to see not only the valleys and lowlands, but also the rising grounds and high hills planted with green olives, whose tops in the time of harvest, are by their weight bent to the very ground. [Juan Bautista Dameto and Vicente Mut, trans. Campbell, 1719]

Introduction: Mallorca in a Spanish context

Through much of the sixteenth century, and especially after Charles V (Charles II of Spain) was made Holy Roman Emperor and King of Spain in 1519, a not wholly unified 'Spain' became integrated into continental Europe. In the Peninsula itself various provinces were united to form what was in effect a new country, a process begun by the *Reyes Catolicos*. Although we have drawn attention to Mallorca's continuing ability to develop strong communication and trade links with a wider world, in essence the island was merely a dot on a large map, even more so as the Spanish Empire spread on the other side of the Atlantic, leaving the Balearic Islands in a more peripheral position. A glance at the index of almost any book in English on early modern Spain will find but few references to Mallorca. [1] As far as trade was concerned – and this became increasingly important in the second of these two centuries – Mallorca was both local and cosmopolitan. Were changes in its landscape in the sixteenth and seventeenth centuries the result of these wider social and economic forces, or were they largely autochthonous (Ringrose, 1996: 198)? While the sixteenth century for Mallorca is sometimes described as being one of 'crisis' in terms of production, the actual spatial organisation of landholding and land-use was beginning to undergo a significant, if rather slow, transformation towards a more capitalist and somewhat less feudal model. It was perhaps only in the context of Turkish expansion in the Western Mediterranean, as the Barbary republics or regencies, that Mallorca figures in the Spanish political–historiographic imagination; in the mid-sixteenth century Philip II even wanted Menorca's population evacuated to Mallorca, a sacrifice of territory at the frontier with Islam (Braudel, 1992: 116). For trading purposes our island was still a significant hinge between Europe and North Africa.

This external factor hindered development in Mallorca in this period. The threat of the Turkish fleet, especially after it was commanded by Barbarossa and other Barbary leaders, meant that much effort had to go into the island's

Figure 8.1 The atalaya of Cala Pí (Llucmajor). One of many built around the coast of Mallorca to give warnings of corsair and pirate invasions in the 16th century. Source: Estop.

defence, especially strengthening key strongholds such as Canyamel, Santueri and Capdepera. For Mallorcans there was an ever-present fear of capture by corsairs, followed by enslavement in North Africa. Improving the system of coastal watch-towers (*atalayas*) first developed in the medieval era (Fig. 8. 1). In sixteenth century Manacor, for example, attacks by Barbary corsairs and the fear of such attacks led to complex arrangements for the warning and defence of the area. Towers along the coast were increasingly used as beacons to warn of attacks, and as places of refuge such as Son Fortesa (Fig. 8. 2), s'Espinagar and s'Enagistes on the road from the *cales* to the town. Most of the large houses of the *possessions* or landed estates were fortified, such as Rafal Pudent. Many of these towers – some round, some square – remain prominent in today's landscape (Rosselló Vaquer and Vaquer Bennasar, 1991). Despite Barbarossa's death in 1546 raids continued – on Pollença, Alcúdia (Cala del Pinar), Valldemossa (Cala d'en Claret), Sant Elm and Andratx in succes-sive years from 1550, and on Sóller in 1561. It was not until the Turks were defeated at the sea battle of Lepanto in 1571 that the threats – and the resulting diversion from development – were reduced. Even then there were major raids on Andratx in 1571, Sóller in 1578, Pollença in 1580 and the island of Cabrera in 1583.

Figure 8.2 Son Fortesa (Manacor) – a fortified *possessió* located near the east coast dating from the 17th century. Source: Volker Stalmann.

Of internal conflicts, the *Revolta Forana* (1450–52) and *Germanies* (1521–23) were to have longstanding effects. They resulted from a massive alienation of rural population from the land – largely occasioned by the loss of communal land to sheep ranching, part of a continuing series of 'town versus country' conflicts. Their initial effect was to reduce investment in agriculture, slow down the rate of land colonisation, reduce the output of bread grains, and generally to sap the island's economic energy; negative characteristics that had to be addressed, as we shall show below.

Demographically Spain continued to suffer from its low density of population, losses to soldiery and empire, and ravage by plagues. The decision in 1609 to expel the Moriscos further reduced numbers that represented both labour and demand for goods and services. In addition, the country suffered from the poverty of its environment: its thin soils and its droughts. The first thirty years of the 1600s were characterised by severe water shortages. The early seventeenth century was one of the driest periods in Spanish history. In January 1613 Bishop Bauzá of Mallorca blessed the land and called upon God to send rain. A religious rain procession took place every other year between 1601 and 1650 (Peterson, 1979: 63–4).

Calculation of sixteenth and seventeenth century population is notoriously difficult, but one estimate shows Spain's population rising slowly from 4.7 million in 1534 to 6.6 million in 1591; but by the mid-1700s it had probably fallen to a little over 4.8 million. It was not to reach the 10 million mark until the last quarter of the eighteenth century (Casey, 1999: 21). Population in Mallorca, on the other hand, in the mid-sixteenth century rose at a more rapid rate than in Spain as a whole, but by the beginning of the next century settled at about 106,000, a figure that remained stubbornly constant through most of the 1600s or even declined (Moll i Blanes *et al.*, 1983: 88). In comparison to the Peninsula, Mallorcan densities were high at around 27–30 per square kilometre, much more like other Mediterranean islands. Sicily, for example, had a density of nearly 40 per square kilometre. Above all, there had been only a slow recovery of population – and thereby labour supply – from the Black Death of the fourteenth century.

Earlier we indicated that this helped shift the agricultural economy towards the less labour-intensive sheep latifundia, but by the early 1500s there was a growing necessity for Mallorca to increase its wheat supplies for its growing population. Throughout this and the next century Mallorca's age-old struggle to feed itself continued. For too long not enough new land had been brought under the plough, and the value of wool fell, and woollen cloth exports decreased, thanks to competition from elsewhere. By the seventeenth century Mallorca, like the rest of Spain, was caught up in the 'great inflation'. This was primarily the result of the squandering of Spain's wealth from its American Empire: why make goods, why produce food, when they can be bought with such immense wealth? ('*Wealth is not so good as work, nor riches so good as earning*', Landes, 1998: 172). The impact of this on Mallorca may well have been less than in Andalucía, for example, but nonetheless it meant that funds did not flow into improving the island's agricultural productivity as strongly as they might otherwise have done, increasing reliance on imports once more.

However, it is important to remember that in the context of Spain as a whole, the transition from Habsburg rule to the Bourbons after 1705 was by no means smooth. The decline under Charles II witnessed further economic and administrative setbacks and the failure to benefit fully from the New World Empire. The Europe-wide War of Spanish Succession (1701–14) had a major geographical focus in the western Mediterranean basin, and in the Balearic Islands in particular, with Menorca changing hands between British and French forces; but Mallorca was fortunate in not suffering directly the ravages of war experienced in the Peninsula; of course, exports were severely affected at times. Politically the Balearic Islands were increasingly drawn into the mainland's realm as a more centralised Spanish state emerged, with a consequent loss of local economic decision-making.

Agriculture, land use and landscape

It has to be remembered that despite the advances since the Black Death, much of the land surface of Mallorca remained uncultivated – not strictly 'wilderness' but where the imprint of society had often been minimal. Bare rock uplands, almost impenetrable garriga, dense and scattered woodlands and heathlands covered large areas awaiting the axe, the plough and fire. Even today only about half of the island's area might be classified as potentially productive land.

What was set in motion in the sixteenth century, and was to accelerate in the next two, were attempts to shift the margins of production and to move away from extensive sheep farming controlled by the old aristocracy. Gradually many of the massive landed estates that had their origins in the feudal system were broken up into smaller, though still large, units, and bread grain production increased at the expense of sheep ranching. These agricultural changes were only made possible by a new form of social organisation in the countryside and the emergence of a new breed of what might be called a 'landed gentry' with an urban background, under-pinned by a raft of farm managers and smaller tenant farmers able to acquire access to lands through new rental agreements. They succeeded in inserting themselves between the large landowning class that had feudal origins and the landless peasant labourers tied to the soil. Large estates still dominated the scene, and many of them remained inefficient, with their owners residing in Palma, or possibly a nearby town if it was considered sophisticated enough. [2] The solution was to divide the large estates into smaller, leaseable units. As rents on these relatively short leases had to be paid in 'cash and kind' there was an incentive to increase productivity in order to make satisfactory profits. This was accomplished in a variety of ways, most notably by the introduction of a new rotation system, by the expansion of new products, especially vines and olives, and by taking into cultivation new lands from the garriga and other marginal areas that still covered much of the lowlands. Thus, new forms of more commercial farming were emerging from the late 1400s (Jover Avellà, 2001). The next two centuries saw more of these huge areas broken up into smaller units, with wheat and cereals on the central plain and olive cul-tivation in the mountains and uplands. In addition, legumes began to appear as part of new rotations, possibly on 'a grand scale' (Tabak, 2008: 172). The most obvious change in the rural landscape to emerge from this set of processes was the building of new large houses on the estates of these new 'farmers': the *posses-sions*, perhaps the most obvious and most dramatic element in the contemporary rural landscape of Mallorca, and worthy of separate treatment (see below) (Jover Avellà and Morey, 2003). In addition to the division of the old feudal estates into new units, some members of the landed nobility let part of their land and the *casas* attached to their 'big house' to *amos*, tenants who were in effect farm managers. They too were often more commercially minded than their noble or urban land-lords. Nonetheless, farming remained polarised between very large and very small units. The accounts of landowners for tax purposes of 1578 (the *estims*) show a wide range in the numbers of proprietors. For example, in Puigpunyent there were only 28, whereas Felanitx recorded 690. In Inca 235 of the landowners occupied 71% of the area, but 1167 holdings were less than 5.0 ha in size, covering the remaining 24% of the area (Rosselló Verger, 1981: 21). Many of these must have been barely able to support a family.

However, the new, larger, tenanted farms were the medium through which a considerable increase in productivity was possible. For example, in Manacor, wheat output almost doubled in the sixteenth century, and barley and oats even more so, accomplished by a new rotation system of wheat, barley, oats and a fallow year (Rosselló Vaquer and Vaquer Bennasar, 1991: 109ff). However, there were two variables that were to affect this kind of agricultural landscape: one was a shift in the macroeconomic world order, the other the result of climate change. The first was the rise, after about 1550, of wheat imports into the Mediterranean from the Baltic region. The second was the onset of the Little Ice Age, whose effects were to persist, to a greater or lesser extent, to the mid-nineteenth century, giving more variable, generally wetter, conditions and temperatures between one and one and a half degrees Celsius below the preceding Medieval Optimum (Tabak, 2008: 204–5). Their combined effect was to lead to a reduction in wheat cultivation and to the spread of the tree crop economy, often on the basis of mixed grains/trees cultivation. New arable crops also became more widespread, such as beans, lentils, peas and chickpeas, which were grown as an alternative source of carbohydrates to expensive, and heavily taxed, wheat. Where environmental conditions were suitable other commercial crops were grown such as saffron, and textile plants such as flax and *canyom* (rushes or reeds). Vines were also widely grown, but in the case of Manacor, for example, insufficient to provide for the area's needs.

Taking in more land from the garriga – *rompudes de terre*, or land reclamation – was made more attractive by tax concessions. It was undertaken by *roters*, who today are often seen as unsung heroes of the Mallorcan countryside, the great transformers of land, converting the old enemy of the land – stones – into a friend (García-Delgado Segués, 1998: 59). In reality, he was a traditional form of labourer living rough on the land he was paid to clear of garriga and stones, usually for a period of nine to twenty-one years, during which time he would pay no rent but would be obliged to clear land and enclose fields with stone walls. The *roters* had *'una imatge de pagès miserable, enfrontat a una terra pobre i amb unes clàusules d'arrendament molt oneroses pel fruit que en podia obtenir'* (Barceló Crespí *et al.*, 1997: 40) ['destitute peasants faced with the poorest land and the most onerous tenancy agreements that could be imagined']. The roter and his family lived in simple stone dwellings whose architecture is thought to have descended from the talayot dwellings of the late Bronze Age. Usually square or oblong in plan with a single entrance, these primitive *barraques de roter* (Fig. 8.3) can still be found in the rural landscape (Valero i Marti, 1989: 157–173).

Figure 8.3 *Barraca de roter.*
Source: November Press and
Mallorca Daily Blog.

The drystone walls they constructed made a considerable addition as one of the most significant landscape features of Mallorca, found in many parts of the island. These *tancats* (enclosures) can be difficult to date, as many were made much later after nineteenth century land reforms. They have a variety of construction types and sizes, and are found in the frontier lands of the garriga and woodlands, in dry farming and irrigated areas, and on the plains and in the mountains (Rosselló Verger, 1964: 282–3). They form part of a widespread and historically lengthy pattern of 'parcellisation' based on the *quarterada* (0. 7103 ha) and its subdivisions, the *quarta* (¼ of a quarterada) and the *horta* (¹/₁₆th). Such enclosures often followed pre-existing roads and pathways and newly constructed ones – essentially service roads. Smaller enclosures tended to be found nearer to the settlements – *jardins periurbans*. Their walls were usually 1. 2 to 1. 4 m high and about 0. 8 m wide, enclosing 'fields' roughly rectilinear or trapezoidal in shape. As this stone walling became increasingly necessary to safeguard these new arable areas from the large numbers of sheep and goats that still existed in the sixteenth century, especially in Es Pla, so construction became more sophisticated and passed into the hands of *margers,* who by the eighteenth century had organised themselves into a guild. In Manacor in 1585 there were nearly 32,000 farm animals (11% of Mallorca's total) of which over 24,000 were sheep and 5,500 goats; walls kept beasts out rather than crops in. This process of carpeting much of the island with these enclosures was to continue through suc-ceeding centuries, extending the notion of *enclosed* agricultural spaces as opposed to the often boundary-less landscape of the medieval era. For example, Santanyí's cultivated area consisted of only three very large land holdings (*possessions*) in the fifteenth century (Rosselló Verger, 1964: 284). In the mountains the nobility still held enormous estates, large areas of which went uncultivated, but gradually more productive units were carved out of the hillsides and forests. This came to involve not only the simple act of enwalling a cleared space, but also the erection of a range of small agricultural buildings such as the *curucull,* to house animals and tools. In the mountains this process of 'modernisation' involved the complex business of building new, or restoring old, terraces on slopes of 1:4 (Jover Avellà, 2001). By the nineteenth century the land divisions –*establiments* – were to result in much further and finer graticules of subdivision.

It was probably in this era that many of the toponyms (place names) of the Mallorcan countryside began to be laid down in association with these enclosures. Grimalt and Rodriguez list as many as twenty-one types of generic field name in the coastal area of Manacor, although not all of them will date from the 1600s, many perhaps reflecting contemporary activity. Amongst them are the *sementer* – a large subdivision of a property, usually arable; the *camp* – a sown field, or one with fruit trees; the *marina* – an area dominated by wild olives and *mata*, and the *tancat* – a large, walled enclosure (Grimalt Gelabert and Rodriguez Gomila, 2001). The maintenance of such fields today presents many problems; they may be a defining landscape feature – part of the patrimony – but in many areas of the island, particularly in the central zone, the spaces they enclose have long been abandoned for agriculture. Now they are often the location for the rapidly increasing number of dispersed and isolated houses of a new quasi-rural class (Binimelis Sebastian, 2006: 10,225).

It is fascinating to speculate on the origins of the geometry of so many of these enclosures. Any 1:25,000 map or aerial photograph, particularly of central Mallorca or the *marinas*, shows the predominance of an ordered, often orthogonal landscape, but what techniques were available to landowners or tenants for laying out and measuring the right angles and straight lines of so many kilometres of stone walls and other boundaries? Since 1229, land had always belonged to someone, but now it had greater value as 'property' and its ownership needed committing to paper in the form of maps. Did they resort to the simple instruments inherited from the Roman *gromotores* or land surveyors that we saw at work in Chapter 5? We know that similar techniques were used to lay out the towns and their fields under Jaume II's ordinances of 1300. Later such surveying methods were used by Spanish imperial colonists in large areas of the Americas (Kostof, 1999: 133). But who wielded the instruments – surely not the simple *roters*? If not, was there a cadre of professional surveyors who did? It is likely that simple alidades and circumferentors were used initially. Did they have an equivalent to England's Gunter's Chain, which made measurement and simple triangulation possible (Thompson, 1968: 10–11)? Certainly by the early eighteenth century continental surveyors had sophisticated instruments based on French triangulation methods, using the graphometer rather than the British theodolite and plane table (Hewitt, 2010: 67). This is another example of a technology helping to fashion the landscape. Unfortunately whoever was responsible has left few of the original estate plans or maps in the public domain. Theoretically by the early nineteenth century Spain had adopted the Napoleonic polygonal cadastral system, but it was not until the publication of the *Ley de Medición del Territorio* (Measurement of Territory Act) of 1859 that the first reliable cadastral maps were produced (Palanques and Calvo, 2011).

In the upland areas the terracing of slopes, which had its origins in Muslim farmers' attempts to create irrigated, workable horizontal surfaces, continued (Carbonero Gamundi, 1984: 98). The rise of olive cultivation for large-scale commercial purposes – that is, for oil extraction and export – begins in the late sixteenth century, but expands in the seventeenth and eighteenth (Bibiloni Amengual, 1992). The usual explanation is that Mallorca needed to find a crop that would pay for the growing demand for wheat that the island could not satisfy, what Manera has called '*la compensació oleícola*' (Manera, 2001: 82); oil would buy bread. This presupposes a rising exogenous demand for olive oil that would drive up prices. In the sixteenth century oil quality was low, and it was used mainly for domestic use and by industry – for soap making, for lighting and for lubricating machinery. This was unlikely to be for domestic consumption or the food processing industries alone. New markets were found in the expanding industries of northern and western Europe, particularly in Marseilles, London and Amsterdam. It also had to be a product that had a high value-to-weight ratio so that it could withstand sea-based transport costs, and one with few intervening opportunities such as those deriving from locations in southern France, Italy and the Peninsula.

We have shown that olive cultivation has a long history in the Mallorcan landscape, so that agricultural and extractive technologies were well understood and established by the mid-1600s, but what was required now was a significant step-like increase in output. The commercial tree was probably raised by grafting or budding on *ullastre* or wild olive trees in the first instance, which would have reduced the costs of starting a plantation. If grown from seed, it does not begin to bear reasonable fruit for fifteen years and reaches maturity of production by forty years. Additionally, it is a crop that really only produces sufficient fruit every other year (see Fig. 8. 4). It is difficult to see the olive and its oil as a cash crop responding rapidly to market demands. The 200 year response in the landscape, which is still evident today, could only have been slow to evolve. Also, it would have required new investment in capital equipment for oil extraction and sufficient labour skilled in cultivation, pressing, packing and transportation. Given the fact that demand was largely ex-Mallorca, it is not surprising to find much of the trade in the hands of non-Mallorcans. It would also have required a considerable expansion in the land area under olive trees. Where was this to come from? We know that ecologically the tree can be grown almost anywhere on the island – the wild olive is widespread – but for it to be grown on the lowlands it would have had to outcompete, economically speaking, the expanding wheat and grain lands of the new *possessions*. It is more likely, then, that more 'difficult, marginal' land would have been used for olives. This meant developing the terraced fields on the slopes of the Tramuntana and its margins, part of what Tabak has called 'the new world of the hills' in the Mediterranean (Tabak, 2008: Chapter 5).

Terraces were hardly a new feature of the Mallorcan landscape by the 1550s–1650s, but to accommodate the emerging olive tree agronomy they would have needed improving and modernising: more work for the *margers*. There is evidence that new terraces or at least enclosures on steeper slopes in the uplands

Figure 8.4 Drawing of ancient olive trees. Olives date from at least classical times; some survivors may be up to a thousand years old. The olive became an important tree crop in the 17th century on the mountain possessions of the landed estates. Olives were crushed in massive *tafonas*, the oil extracted exported to the mainland, the Spanish empire, and to the soap makers of southern France. Source: Gaston Vuiller, 1888.

were being constructed at this time. But terracing is a highly labour-intensive and slow process; in previous – medieval – generations much of it had been the product of slave labour. Although the olive oil economy stretched well into the eighteenth century from these early beginnings, it seems more likely that new land uses (olives) were introduced first onto improved, ancient terraces rather than on newly built ones, and in the hill country around Palma, so as to reduce land transport costs to the port (Jover Avellà, 2001: 121). As demand rose, so new terraces were built, often on slopes greater than 1 in 5 on land over 600 m. By 1657 nearly 900,000 *quartans* (3.731 m litres) of oil were being exported from Palma. Some terraces grew various crops according to demand so that those at Banyalbufar were renowned for their viticulture, until the effects of phylloxera were felt towards the end of the nineteenth century; but they had also grown wheat to raise money for taxation (Carbonero Gamundi, 1983).

A new approach to an old crop began, then, to transform the upland landscapes of Mallorca in the sixteenth and seventeenth centuries, while the 'aristocracy of the mountains' made immense fortunes from the trade, much of which was translated into many of the palaces of Palma. In the mountains themselves, when the trade in olive oil began to decline in the late 1700s, the terraces began to fall into desuetude, creating and leaving behind a set of relict features – *safareigs, marjades, bancals*, etc. Many of these had Arab or Berber origins, such as the complex system of terraces and their attendant water distribution system around Banyabulfar, but only a constant programme of renovation and rebuilding over many centuries permitted their continuation (Carbonero Gamundí 1983 and 1984). While being part of the 'romantic' appearance of Mallorca on many tourists' itinerary, today they present enormous problems of conservation. By the second half of the eighteenth century new products and new approaches to the island's grain shortages and growing population began to present themselves, leading to the introduction of new landscape features.

Possessions or landed estates

If the stone walls and terraces of this period were a noteworthy landscape feature, on a much larger scale were the *possessions* themselves, the large – often very large – estates and country houses that are ubiquitous in rural Mallorca. It has been estimated that towards the end of the nineteenth century there were over a thousand in Mallorca. In many cases they may be seen as exemplifying the radical changes to the Mallorcan agriculture that began really to alter the rural landscape and its world in the 200 years from the middle of the sixteenth century. According to Bernat i Roca and Serra i Barceló they became '*a centre polifuncional de gran complexitat on, a més de la producció agrígola pròpiament dita, s'hi donava tota una sèrie d'activitats de transformació de diversa naturalesa, sent una part d'elles clarament manufactureres*' (Bernat i Roca and Serra i Barceló, 2012: 63). The *possessió*, then, was centred on a small group of buildings, the domestic versions of which were known as *les cases,* equivalent to the *lloc* in Menorca, the *mas* or *masia* in Catalunya, the *cortijo* in Andalucía and the *pazo* in Galicia. They had their origins in the fifteenth century, or even earlier in Berber rafals and alquerias, but they were consolidated in the sixteenth century following the rural revolutions mentioned above. Besides houses for *el senyor* and the *amo* (essentially a principal tenant who, in addition to farming in his own right, might act as a foreman for the senyor's part of the *possessió*) there were specialised buildings for grain storage, almonds, olive pressing etc., depending on the type of land use and farming methods. A large estate would have had a chapel (Fig. 8. 5). Many of the older *possessions* were fortified, often with towers such as that at Bàlitx d'Avall in Fornalutx, while others received an improving neoclassical façade with a coat of arms and a sundial, as at Galatzó in Calvià. On the larger estates there may have been as many as 100 workers. In this sense the *possessió* continued the high degree of nucleation to be found in Mallorcan settlement patterns. Although they represented a dispersion away from the medieval settlements, because of their self-contained nature that included workers' housing, there was less dispersion into the fields and woodlands than would have been found amongst freeholders in northern Europe at this time. Generally the principal house stood at the centre of a system of large fields with smaller units of, for example, the *amo*, some distance away. A wide variety of farming landscapes might be observed around the *possessió* ranging from gardens

Figure 8.5 The chapel of the Ruberts *possessió*.

near the house with woodlands, enclosed fields for crops, vineyards and pasture. Towards the periphery of the estate would be the marginal land being worked by the *roter*, representing a moving frontier between cultivated land and the *garriga*. Many of the larger, new and substantial houses at the heart of the *possessió* are thought by some to be based on the Catalan *masia*, but clearly a Mallorcan form has evolved (Fig. 8. 6). Oblong in plan with the roof sloping in two directions from a central beam along the long axis, the main doorway is in the gable end. The building, with vertical stone walls and terracotta roof tiles, is usually divided into three parallel aisles on both storeys. However, a smaller second type, perhaps with its origins in Andalucian/Muslim architectural tradition, also developed. Originally single storey with a monoclinal pitched roof sloping to the front elevation, they have a central door and window openings to either side in the front façade. Later many developed to be double in size, both vertically with two floors, and in depth under a double pitched roof. These simpler houses were built by and for the smaller landowner or tenant lower down the feudal hierarchy. Widespread through the central and southern parts of the island they are particularly numerous in the south and east in Felanitx, Porreres, Santanyí and Manacor (García-Delgado Segués, 1998). Both house types are amongst the most distinctive features of the human geography and rural landscape of Mallorca.

The changes wrought in the late sixteenth century stemmed from two sources: the considerable rise in population numbers from the low points of the fourteenth century Black Death and the rural revolts of the fifteenth and early sixteenth centuries referred to above. The nobility took the opportunity to oust many of the smaller farmers (*pagesos*) and consolidate their holdings into larger, more economic units. The displacement resulted in large-scale rural unemployment and the need to create paid work, ranging from day labourers to olive pickers and a wide range

Figure 8.6 A typical smaller 18th century farm house near Porreres built when lands were further subdivided. Some of the ranges of farm building have been divided into houses in the late 20th century.

of skills required in the new *possessió* and its *cases,* thus creating a new set of social castes (Jover Avellà, 2012: 209–229).

Some extensive examples drawn from various parts of the island will serve to illustrate the changes to size of holdings in the sixteenth and early seventeenth centuries in a little more detail; but note the continuation of very small units alongside some of the modernising, larger ones.

From the Manacor municipality in the east of the island a cadastral survey of 1578 showed that the 927 smaller holdings in the municipality occupied only 2. 75% of the area, 50% of them of less than 5 ha and 2% between 4–9. 9 ha. A tax assessment of about the same date revealed that large land-holdings still dominated, with eight *possessions* valued together at more than 107,400 lliures when the average value of all 108 was only 3524 lliures. Some represented enormous wealth, such as those belonging to Gregori Villalonga, whose three estates – El Rafal Nou, La Clotana and La Cove – were worth at least 18,000 lliures. Data from earlier in the century illustrates the new leasing arrangements, with rent partly paid in cash and part in produce or service. For example, in 1525 Jaume Vives let the *possessió* known as La Blanquera for four years to Jordi Melcior Riera. It had 210 sheep, 4 oxen, 3 bulls, 2 cows in milk and 22 goats. In addition, the owner gave 13 quarteres of wheat for seed. The rent was to be paid for with 32 quarteres of bread wheat and 35 lliures in cash annually. In another example, in 1538 Margalida, widow of Romeu Desclapers, a squire, let the *possessió* known as La Cova to Melcior Riera for 5 years. On the property there were 200 ewes, 28 goats, 3 *segals* (breeding mules), 4 bulls, 3 cows, 4 mares, a mule, 2 donkeys, 15 quarterades of fallow land, 20 quarterades sown with wheat and 10 to be sown, 2 ploughs with yokes, one for mules and one for oxen. Each year the tenant had to give two dozen cream cheeses, 3 lambs and 3 kids, pay for a day's worth of planting fig trees and plant two bushels of beans. One mule was to be kept for going back and forth to town. In addition the rent was set at 45 lliures annually, together with a contribution of 45 quarteres of wheat (Rosselló Vaquer and Vaquer Bennassar, 1991; Vibot, 2006).

From the south-west the Galatzó *possessió* in Calvià was another large holding, again, Muslim in origin, but its first-noted Christian medieval owner was Berenguer Burgues, and from 1341–1400 it passed to Aparici Cirera (Fig. 8. 7). It was divided into three following a marriage settlement, and then the three parts were reunited by Berenguer Vibot. By 1578 it was valued at 16,000 lliures. By the seventeenth century it had added a defensive tower, chapel, wine cellar, oil press and water mill. It cultivated vines, olives, cereals and legumes, with a considerable pastoral element supporting 800 sheep and goats. In 1627 it was sold to the first Count of Santa Maria de Formiguera (1627–1694) – the so-called Bad Count. By 1818 its value was estimated at nearly 90,000 lliures, one of the highest in Mallorca. By then, in addition to the houses and 2 quarterades of gardens, it possessed 19 quarterades of olives and carobs of the first quality, 200 quarterades of second quality and 198 of third. It had 30 quarterades of fields without trees and 550 quarterades of mountain and uncultivated land. It had two water mills valued at 666 lliures (see Fig. 6. 5). By the end of the nineteenth century much of its cultivated area was given over to fruit growing, especially mandarin oranges.

Figure 8.7 The Galatzó *possessió* (Calvià) – see text.

Another example from Manacor was the estate of Es Fangar, near present-day Son Macià. From the fourteenth century it belonged to the Truyols family. Much of their land from S'Horta to Son Fortesa was uninhabited and only slowly converted to agricultural uses, with eight or nine of the *possessions* created that we see today. In 1578 Nicolau Truyols let the principal possession for 5 years to Pere Nadal and his son. There were 1200 sheep, 1700 goats, 3 donkeys, a mule and two breeding mules, a horse, 15 cattle and two cows. The rent in kind consisted of kids, piglets and fat pigs, 90 quarteres of wheat, 3 quintars of cheese, etc.

In the mountains a mixture of mini- and latifundia also prevailed, but the agricultural economy was different in this kind of environment, a landscape that often exhibited only '*un sello de vida en las frecuentemente desolados paisajes*' ('little sign of life in a frequently wasted landscape') (Ferrer Flórez, 1983: 15). Cereals were much less noticeable except in the wider valley bottoms, and where grains were grown they were as part of a four-part rotation: fallow, wheat, *restoble* (barley and oats) and pasture. As in the lowlands, this tended to increase through the 1600s as division of old feudal lands increased. For example, the old latifundium of Massenella was divided into one large estate and three smaller ones. The ancient hacienda of Canet was split into four discrete olive plantations and let out to tenants. All these new estates had to be equipped with houses for the new tenants and their workers, new storage facilities for the crop, and buildings for tools and equipment. Many of the old properties descended from rafals and alquerias, however, remained unreformed and still concentrated on sheep and forestry, often in units of over 1000 ha. Population numbers doubled in the mountain municipalities, but actual numbers remained small. Some areas were highly nucleated, such as the Pollença district, whilst the more accidented relief of, for example, Escorca, meant it retained a dispersed population. Olives were the principal crops on the steeper slopes and the terraces such as in the Fornalutx *barranc*. In the late seventeenth and early eighteenth centuries there was a brief revival of this wood/pasture economy as a rise in demand for wool led to more sheep on the mountain pastures, charcoal output was increased for the metal-working industries, acorn pastures fed larger numbers of pigs, and more land was enclosed for animals.

The *possession* then emerged in the 200 years after 1500 as a distinct landscape feature, representing a new economic and social force in the countryside, setting the tone for further advances in agriculture and land use that were to be made in the next century, whether in the plains or the mountains and hills (Jover Avellà, 2001).

After 1715 the larger units continued to pass to this new class of landed gentry, many of whom had professional origins in Palma. These new proprietors were more innovative, increasing the cultivated area, introducing new rotations leading to a new kind of latifundia characterised by a commercial approach to production serving local and distant markets, even export. Vines and cereals were grown in the lowlands, olives in the uplands.

The changing rural landscape after 1750

So far we have charted the changing rural scene from the sixteenth century, but from about the middle of the 1750s the agricultural economy began to shift once more, into what Carles Manera has called *'la irrupció d'una agricultura alternativa al latifundi'* (Manera, 2001: 37), a process made possible by a variety of factors including the new leasing arrangements, the development of the *possessions* and the taking in and enclosure of new lands. In the next 100 years to about the mid-nineteenth century further shifts can be observed that involved the expansion of new crops and a decline in dependence on the olive.

Olive production was clearly used as the income stream with which Mallorca purchased its grain imports – the island produced only 155,000 quarteres of wheat in the 1720s but required 400,000 to feed its growing population. The monarchy was in constant fear that food shortages might trigger social unrest of the kind experienced in the early 1500s. The amount of grain that could be imported was largely determined by the price of olive oil; oil and wheat prices had to keep pace with each other, especially in the longer run. If oil output was to be increased, then the extent to which growers were prepared to expand the area under the crop, either by taking in new land such as by terracing, or by substituting for the land under other crops, was price dependent. In this way economics could influence landscape. The large oil estates were originally under the control of the landed aristocracy, who were often negligent of their holdings and knew little of the intricacies of trade. The new leasing arrangements made it possible for the Palma-based merchant-traders to become producers, enabling them to regulate supply.

From the mid-seventeenth century the farming system below the level of the large landed estates had already begun to diversify away from the oil/cereals agricultural economy. On the smaller farms, many of which had been carved out of the old feudal estates under the new tenancy agreements, new crops were introduced, mostly based on tree crops such as carobs, almonds, citrus fruits, legumes, vines and pig rearing based on figs. The table below shows the changing values of these crops relative to the oil and cereals:

Table 8. 1 Crops 1784–1820/35. Percentages of output by value at 1835 prices.

	Cereals	Legumes	Oil	Wine	Almonds	Carobs	Figs
1784	53	9	31	8	–	–	–
1820–35	52	13	19	9	4	2	1

Source: Manera, 2001: 113.

As the nineteenth century progressed, this pattern was reflected in the area under a variety of these new crops:

Table 8. 2 Crops 1818–60 (hectares).

	Cereals	Legumes	Oil	Wine	Almonds	Carobs	Figs
1818	166271	–	43698	19786	–	–	–
1860	122789	12340	25949	15543	5961	7660	12789

Source: Manera, 2001: 113.

New elements were being added to the Mallorcan landscape that were to persist through to the present day – the black carob bean known to the English visitor as St John's bread, the February sea of almond blossom, the soft brown and purple fruits of the fig, the familiar oranges and lemons, and the vines of central Mallorca that support a growing wine industry. These, to a greater or lesser degree, were to perform a valuable function in giving the island new markets for export, helping to address the earlier *situació dels cereals caòtica i pràcticament irresoluble* (Manera, 2001: 53) (a chaotic cereals situation, practically unresolvable'). We shall deal with them in more detail in the next chapter.

Landscape with trees

In Britain in the eighteenth century trees became a major topic of interest to the larger landowners, who sought to empark the land immediately surrounding their country houses, following the new principles of landscape gardening set out by Capability Brown, Humphrey Repton and others. In Mallorca, although the effects of this kind of aesthetic movement were felt by some, the transformation of large parts of the countryside was much less. To the new *rentiers* of large estates their recently acquired lands had to be put to more productive and less 'landscaped' uses.

The introduction of tree crops into the story of Mallorca's landscape, noted above, is an opportune moment to come back to what might be called the native stock of trees in the 1700s. By this time woodland was dominated by pine and holm oak with black and white poplar and lesser amounts of elm, walnut and ash. The geographical distribution of stands of more or less continuous trees was limited to the upland areas, especially for holm oaks, while pines were to be found more widely, especially on the coastal flats. A third location was the often intermittent water courses across parts of the island whose sinuous shapes can still be detected. The 'timber' in these kinds of woodlands gave a number of valuable resources including charcoal, construction and timbers for small ships, logs for household and industrial fuel (including lime burning) and turpentine. Throughout this period there was often concern that the tree population would be reduced beyond its reproductive capacity. While some areas and species did show that, a glance at the contemporary Mallorcan landscape demonstrates both the recuperative power of trees and the fact that the stock was more carefully husbanded. However, there probably was some link between the nearly tripling of human population numbers in the 300 years from the end of the fifteenth century, leading inevitably to increased demand for wood in all its guises. In the longer run we know that the area under pine and holm oak actually fell from 35,300 ha in 1748 to 15,256 ha in 1881, although much of that must have been in the nineteenth century (Berbiela Mingot, 2010: 551).

At a time when more enlightened economic ideas were taking hold in the eighteenth century, woodlands and forests were increasingly likely to be seen as important resources that had to be managed, not laid waste. At one level alternative uses would be determined by the rate of return on cleared land, and when prices for figs, olive oil and carob beans were high some clearance would have taken place, especially on 'better' land. Of course, such prices did fluctuate so that reversion to woodland also occurred, and much 'monte' emerged as a mixed land-use of crops, grazing and forestry. Woodland in Mallorca was also to be found on communal lands, an important resource for the less well off; but the take-over of these areas by enclosures and parcellisation from as early as the sixteenth century meant that some of it was lost (Brunet Estarellas, 1991). The removal of trees was determined by clear rules as to which trees could be felled and which pruned or brashed, particularly in charcoal-making areas.

One agency in Spain that is often accused of deforesting much of the country-side was the navy; but surely until the coming of iron- and steel-hulled ships, the Armada was much more likely to conserve and manage timber resources rather than destroy them. In 1748 it carried out an inventory of the trees in Mallorca. Someone (Pedro de Hardeñana) actually counted all 7,186,710 of them – 4. 7 million pines and 2. 4 million holm oaks. Grove and Rackham have mapped these data showing a concentration of pines in the coastal areas such as Muro with 1. 3 million pines and Escorca with 1. 4 million holm oaks (Fig. 8. 8). The purpose of the navy's survey was to estimate resources for the revival of the fleet.

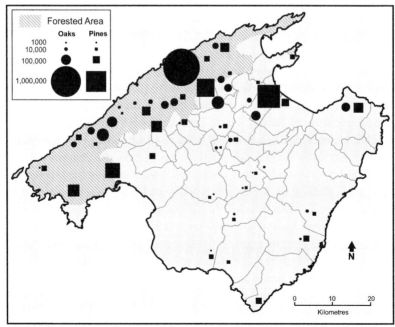

Figure 8.8 Map of woodlands in the late 18th century. Source: adapted from Grove and Rackham.

As industries and population grew there was an inevitable increase in the demand for timber products. For example, in 1767 Mallorca produced 3. 6 million litres of olive oil, 90% destined for export; this had to be shipped in wooden barrels, as did figs and almonds. The soap-making industry was expanded; this and iron-making were major consumers of fuel wood and charcoal. The value of wood output on tenanted lands was, like any other crop, shared between tenant and landowner. Timber was widely used in construction, although few of Mallorca's trees grew tall enough for large projects. Even for shipbuilding many of the timbers were unsuitable, leading to timber imports from an early date. Nonetheless, by the late eighteenth century there were over 700 *fusters* at work in Mallorca, nearly half of them located in Palma engaged in making carts and wagons, furniture, house building, shipbuilding and cooperage, although many of the latter were to be found in Felanitx and Binissalem (Escartín *et al.*, 1996: 22). Sansó has identified 566 *fusters* located across thirteen towns in 1784, 61% working in Palma (Sansó Barceló, 2011: 22). It was in the later part of the next century that charcoal making began to exert much greater pressure on forest resources. Perhaps we should remind ourselves of Oliver Rackham's comment on the exploitation of English woodlands, namely that it was highly unlikely that industrial consumption was a major cause of their destruction, 'for trees grow again, and a wood need no more be destroyed by felling than a meadow is destroyed by cutting a crop of hay' (Rackham, 1976: 91–2). In the growing rationality of the eighteenth century Enlightenment it is unlikely that some kind of economic death wish would have been perpetrated on Mallorca's woodlands. Since then the island's holm oak cover has more than doubled, and its pine woods have increased by a factor of at least five.

An urban renaissance in the seventeenth and eighteenth centuries

If change was the order of the day in the rural areas, then in medieval Ciutat and the smaller towns new developments were also taking place, particularly in Palma, where a new Bourbon aristocracy – the *'botifarres'* – was converting its new-found wealth into city palaces and dominant public buildings, many of which were associated with the Church and religious houses. In the *part forana* the towns founded under Jaume II's ordinances were being consolidated, with many of the *illetes* being built on for the first time so that they began to take on the air of proper urban places rather than simply villages.

Two forces affected the urban morphology of Palma in this era – the building of a new set of walls and the diversion of the intermittent River Riera (Tous Meliá 2009: 51). On the ground today these two features are not so obvious, thanks to later buildings and road developments, but they are at once evident in Garau's map of 1644 (Fig. 8. 9). The new walls date from the late 1500s, and their structure reflected ideas on military defence and the high degree of political turbulence and warfare in the Western Mediterranean at that time (Tous Meliá, 2004). The major technological innovation in the military sphere was artillery capable of destroying

Figure 8.9 Garau's map of Mallorca, 1644.

previously impermeable city walls, and from the reverse perspective, of defending them. Philip II appointed the Italian military engineer Giacomo Palearo to design this third set of walls. He and his successor – his brother Giorgio – planned for a set of straight curtain walls punctuated by a series of ten diamond-shaped bulwarks so that raking fire could be directed parallel to the walls. This eventually gave the zig-zag pattern so common to such Renaissance defence systems in many European countries. Today this pattern has been frozen in the morphology of the *avingudes* (*avenidas*), laid down when the walls were demolished at the beginning of the twentieth century. The building of these walls and six new gates took most of the seventeenth century to complete, with the sea wall and its bastions not started until the early part of the next century. They were built just beyond the line of the second set of medieval walls, which created new opportunities for building or new open spaces between them. The positioning of the gates at this time was to determine much of the shape of the road network in the modern era, and thereby the location of the later suburbs (see Fig. 7. 8).

The Riera, as we saw earlier, had a regime typical of most Mediterranean rivers, that is, in the summer months it was primarily a dry bed with water flowing only in the winter. However, flooding on a drastic scale took place from time to time (1404, 1443 and even in 1618 before the diversion was complete), again typical, the consequences of which were often disastrous for the medieval morphology of the 'low town'. When flowing, the river had its uses as a waste disposal system and as a polluted water source. When dry, the river bed acted as the typical Spanish *rambla*, a scene for the *paseo*, part of the 'city as theatre', common in the 1700s and 1800s. On balance it was judged to be nuisance at best, a danger at worst. A decision was taken to divert it in 1613, down the west side of the new walls. This entailed building new bridges across it to take new roads leading from the new gates. Its former channel led to two new elements in the townscape, Es Borne and the Rambla. The first of these developed as a formal construction based on its original pedestrian function, paved and lined with trees, but providing new opportunities for building alongside, such as Can Morell (Solleric), perhaps the grandest of the señorial houses of the eighteenth century, and paid for by the family's tremendous olive oil fortune. The Rambla was less well developed at this time and remained a venue for festivities and display. New buildings were also sited here, such as the church and convent of Santa Teresa de Jesus, built for an enclosed order of nuns between 1624 and 1688.

All in all, the period between the end of the 1580s and the 1750s was the most formative in architectural terms in Mallorca's history, an urban renaissance in which new styles of architecture were introduced. Baroque became the most common style, most evident in the addition of lavish doorways and casements to most of the churches, such as Montesión (1683) and the Convent of the Conception (*c.* 1630). In Palma many of the town houses and mansions are still very evident in the townscape, now forming part of the tourists' itinerary as they search out the palaces and the patios. These were built throughout this period, and on an enormous scale. Unlike the town mansions of the aristocracy in London or Berlin at this time, these palaces were built into the streetscape as part of a continuous terrace rather than freestanding. With a largely blank façade pierced by an impressive wooden gateway leading into the familiar patio, they present a forbidding but powerful exterior, emphasised by their wide, overhanging and intricately carved cornices. A large stone staircase then leads upwards to the extravagant living quarters above. There are many examples of this type of building in Palma, but among those that exhibit best the architectural principles of the early part of this era are Can Oleza in Calle Morey (Fig. 8. 10) and Can Catlar in Calle de Sol (Lucena *at al.,* 1997). Later in the mid-eighteenth century the architectural style became the more severe Classical, and the aristocracy was not slow to adopt it, as seen in Can Vibot in Calle de Can Savella, Can Berga built in the bend of the former Riera channel at Plaza del Mercat and Solleric, mentioned above. Similarly, the churches went in for its simplicity too, often as a refaçading of an older building. The most important public building of this time was the Ajuntament (1649–1680) with its huge cantilevered cornice, facing the Plaça de Cort, another open space for public display. This town hall was, and remains, indicative of increasing municipal authority, a symbol in stone of the city's wealth.

However, the street plan within which these new forms were built remained very much as it had done in medieval or even Muslim times. Plots might be altered and amalgamated through mergers in order to accommodate these new opulent buildings, but the road network remained inviolate. It was not until the modern period that town planning would alter the movement infrastructure of the city, largely in the service of traffic. The narrowness of the medieval streets led to

Figure 8.10 C'an Oleza in Carrer den Morey. An unassuming façade fronts directly onto the street, but the patio reveals one of the prestigious town houses originally built in the 16th century but refurbished in the baroque style in the late 17th century. From a postcard of 1910. Source: Muntaner.

vertical developments on a new scale, increasing overall building densities, something that was also fashioned by the constraints imposed by Palma being a walled city. The city's population rose from 25,988 in 1667 to 33,121 in 1746, but fell slightly to 31,965 in 1784, while rising again to 36,617 by 1825, always between 23 and 26% of the island's total (Manera, 2001: 71). These increases meant not only rising density, but also led to a marked separation in the social spaces of the city between the opulent *senyors* and merchants and the poorer labouring and servant classes. Living conditions in the narrow alleyways must have been cramped, ill-lit and dangerously unhealthy for the poorer sections of society. Plague, especially cholera, was a common visitor to Palma throughout the early modern period, and the environmental conditions that encouraged it were not seriously addressed until the mid-nineteenth century.

These kinds of changes were replicated in the country towns of Mallorca. Mostly located inland, they did not have the same strategic significance as Palma, although there was a residual fear of attacks by corsairs. It appears that no new walls for such towns were ordered for this disturbed period, but those of Alcúdia and Capdepera were strengthened. The main features of town development were simply those associated with growth – additional housing, improved streets and better water supply. By the 1797 census there were seven towns with a population of more than 5000, indicative of a growing urbanisation of the population (Fig. 8. 11). Gerónimo de Bernard's tour around the towns of Mallorca of 1789 is a

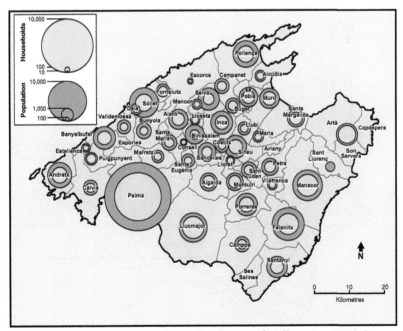

Figure 8.11 Map showing population distribution of Mallorca in *c.*1786. Source: constructed from tabulated data in Vargas Ponce, 1786 (Ed. Moll i Blanes, 1983).

useful insight into urban development towards the end of the eighteenth century (De Berard, 1789). The town plans of Jaume II, dating from the early 1300s, were at last being fulfilled more rapidly – to take one example, that of Manacor, Mallorca's second town in size by the early eighteenth century. By the beginning of the 1600s over 500 houses were recorded in the town, slowly filling the *illetes* of the original ordinances to form a compact, nucleated settlement mainly to the south of the hill upon which the principal church stands (Ferrer Febrer and Carvarjal Mesquida, 2003). As De Berard's map of 1789 (Figure 8. 12) clearly shows, there was an additional consolidation of building along the axial roads coming into the town and in the spaces between these roads, but even by the early 1900s Manacor remained essentially a planned medieval town, awaiting the impact of industrialisation for future growth to take place. The municipality's population reached almost 6000 the late 1700s. Thanks to owning or leasing some of Mallorca's largest *possessions* in the surrounding countryside, the local gentry were able to afford houses in the town as well as places in Palma. With a growing number of crafts based in the town, including textiles, and the spread of viticulture in the eighteenth century, new wealth and trades were brought to the town, reflected in the opening of new workshops. By 1800 these included 13 metalworking shops, 50 linen and woollen mills, 16 skinners, 36 woodwork shops, 20 shoemakers, 10 brick-making plants, 12 quarries – altogether employing 328 men. In addition there were 22 distilleries in the town (Morey Tous,

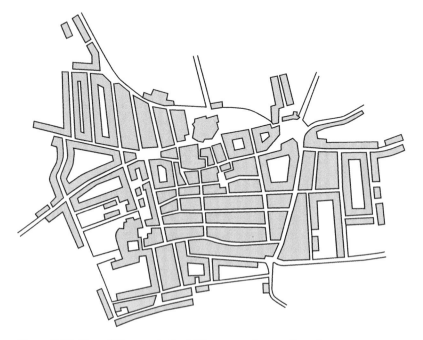

Figure 8.12 Map of Manacor in the 18th century. Source: Berard.

2001). Convents and monasteries featured as noteworthy elements of the urban morphology until the 1830s, when their lands were broken up and auctioned for development. The churches and sometimes the cloisters remained, as at St. Vicenc Ferrer's, whose beautiful conventual ranges date from the late seventeenth and early eighteenth centuries and now house offices of the *ajuntament* and the town's library, having been taken over by the State as part of the disentailment of 1835 (Roman Quetglas, 2001).

By contrast, Andratx in the south-west of the island was not one of the towns of the *ordinacions* of 1300, and unlike most settlements was located close to the coast, leaving it exposed to frequent attacks from the sea by North African corsairs ('moors'). Thus it needed defending by lookout towers and walls. Its morphology consisted of a single street (Calle Major) with its parish church on an eminence at its eastern end and minor lanes running down to the port (Fig. 8.11). Berard records 278 houses in the town in 1789. Its hinterland supported olives, carobs, vines (providing *vino exquisito*) and arable crops of wheat, oats and barley processed by six water mills and seven windmills. Six major *possessions* dating from the break-up of the lands of the medieval Bishop of Barcelona and before dominate the surrounding landscape of river terraces and broken hill lands (De Berard, 1789: 15–25). As with so many places in the rural interior, there was a close economic symbiosis between the town and the surrounding countryside.

These two centuries can be seen, then, as a transitional era between the truly medieval and the modern. As new forms of economy evolved in the countryside and in the towns – some the result of national and international forces, some coming from more certainly self-contained forces from within Mallorca – so the landscape started to ease itself away from the huge estates of the aristocracy to the

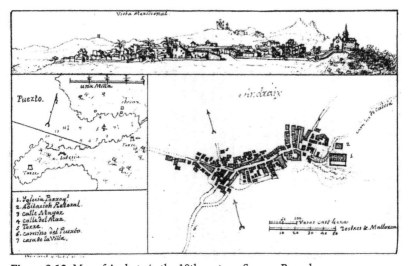

Figure 8.13 Map of Andratx in the 18th century. Source: Berard.

more controlled scenery of the nobility and the rising liberal classes, a landscape that increasingly showed signs of the capitalist transformation of island society. The small farmers and landless peasants at the bottom of the hierarchy played their part in this transformation process through their considerable and poorly rewarded labours, but their share of it was to wait until the second half of the nineteenth century.

9

The long nineteenth century, 1820–1920: the beginning of modernisation

Mallorca and the changing Spanish space-economy

The decline of the old regime in Spain was a slow and painful process involving loss of empire and protracted economic development. The transition to a more modern economy was so long-winded that it might be said to have been reached only by the 1980s with the country's accession to the European Union. To those raised in north-western Europe, and especially in Great Britain, West Germany, northern France and Belgium, the nineteenth century is a term that signifies a landscape based on industrialisation and urbanisation, especially if the 'century' is extended from about 1760 to the First World War. It was the long period when the population began to leave the countryside for a life in cities and towns and population numbers rose rapidly; a life dominated by the discipline of the factory system and the mine company's whistle, a world in which the poverty of the peasant's cottage was exchanged for the drabness of the back-to-back terraced house, from a closed world of limited geographical horizons to one of high connectivity via canals and railways, from a world of squirarchical domination to one of slowly increasing democracy.

It is a familiar picture, but was it the experience of Spain and her Balearic archipelago? An earlier generation of historians sought to show that the Peninsula and the islands were not part of the revolutionary changes in economy and society that swept through more northerly parts of Europe, that the landscape, described in the previous chapters, of latifundia, of low population densities, of slow urban growth and of poor internal communications persisted for most of the following century. Spain was not without raw materials – it produced half of Europe's copper, nearly two-fifths of its lead and about a seventh of its iron ore in the thirty years after 1880. Thanks to phosphates from North Africa, it became a major manufacturer of fertilisers, but nearly all of these extractive and primary processing industries were dominated by foreign capital. Spain remained, it was said, isolated, and without its empire – after 1824 only Cuba, Puerto Rico and the Philippines were left from the Golden Age – it had few markets but its own internal ones. Indeed, it was the low level of demand that held back Spanish industrialisation more than anything else. Eventually, population grew markedly in the nineteenth century, but from a very low base, and its spending power remained restricted, with no real energy in national consumption. The landscape was said to have a timeless quality that would go on and on; to some, the romantic but backward realm of Eternal Spain, *La Espanya profunda*.

Of course, this view was never entirely accurate, but it served a particular purpose. In fact, as in any other country, there were marked spatial variations in the experience of 'modernisation'. The society and economy of the whole of Spain may not have been transformed in the nineteenth century, but parts of the country certainly were, especially in the Basque country and in Catalunya and, when the microscope is properly focused, in Mallorca too. And as we saw in the previous chapter, this process began at least in the 1700s, so that the geography of change was far from ubiquitous.

In Mallorca the agricultural reforms were hardly 'revolutionary' in their early years. There was no real enclosure movement backed by legislation as we knew it in England (although the reforms that began in the 1830s were significant); no large-scale rural depopulation, no marked increase in agricultural productivity, despite certain shifts in land-use; and there was little, if any, mechanisation of farming. By European standards grain yields remained low and total output insufficient to feed the island's population. However, although longer-drawn out, changes did take place in the countryside that subtly affected the landscape. Much of the period to the middle of the nineteenth century might be best described as a period of transition in which the changes observed in the previous 100 years were continued. The Mediterranean trilogy of wheat, olives and vines continued, but with more of the production finding its way into primary processing industries such as flour milling, soap making and distilling. The land reforms after about 1860 led to a new rural landscape of small farms, fields, and an expanded arboriculture of relatively new tree crops, especially almonds, figs and carobs, which themselves produced raw materials for the food processing industries. In this sphere Mallorca was better connected with agricultural improvements in Europe, but more particularly, the island had an active, articulate merchant class well connected to emerging markets on the Continent and in the Americas, acting as catalysts and links between Mallorca's farmer-producers and their markets overseas.

In the industrial sector Mallorca further expanded its textile industry, especially cotton textiles, a new material that was not encumbered with the 'old ways' associated with the 'ancien régime' of wool and the leather-working industries, particularly boot and shoe-making. Their growth awaited the introduction of steam power and a factory system to replace a large part of the craft-based cottage industries that had been in existence since the later Middle Ages. Towards the end of this period newer, small factory-based activities such as furniture, woodworking and agricultural machinery developed, later associated with the availability of electric power. For many of these emergent industries the island depended on imported raw materials. Sales were largely restricted to Mallorca's home market, itself a function of population growth, and to the remains of the once huge Empire. Mallorca possessed few mineral resources – metals and carbons, little in the way of reliable water power sites and poor internal transportation networks. In addition there was only a very limited social infrastructure, with low levels of education; by 1900 56% of Spain's population was illiterate; the British figure was 5%. There was little or no technical training and few intellectual sources such as a university to stimulate invention and innovation. Science and technology were much less highly rated than the humanities, theology and the law.

Despite these apparent drawbacks, research has begun to expose a somewhat different picture of Mallorcan industrialisation (Manera, 2001, 2006; Roca Avellà, 2006). For the industries that did develop there must have been entrepreneurship and sources of capital, even if parts of each may have originated from elsewhere, such as from Catalunya. Mallorca, as in previous centuries, was not as isolated from the economic and scientific thought emanating from north-western Europe in the late eighteenth and early nineteenth centuries as might be supposed. During this period the island had a rather more dynamic economy, in touch with developments in Britain, France and Germany. The primary processing industries mentioned may have been constrained initially by certain institutional factors such as the strength of the old guilds, but by the middle of the century considerable progress had been made. The old craft industries did produce some surplus capital; the resistance of the guilds was largely overcome by developing factories away from Palma; new machinery and steam power came in from outside, and home consumption grew. Industrialisation may not have followed the classic British model of cotton, iron and steel, canals and railways and booming provincial cities, but local writers are now able to refer to '*una industrialització sense revolució industrial*' (Roca Avellà, 2006: 10) or '*manufactures sin revolució industrial*' (Manera Erbina and Petrus Bey, 1991: 13–54). However, it would appear that the main purpose of much of this writing is to show that Mallorca was not isolated from the main trends in European economic development, even if much of it was at a slow rate of 'progress' (non-revolutionary) and occurring somewhat later than elsewhere. It also seems tied up with the *political* notion so important in Spain, of weighting regions (the Balearics, in this case) more heavily than the nation (Spain).

For an island, trade remained paramount. Mallorca and Palma illustrates quite well the apparent paradox that overseas business was both 'local' and 'cosmopolitan'. Palma was primarily part of a network of west Mediterranean port cities that included Genoa, Marseilles, Barcelona, Malagà and the ports of North Africa, but it also traded with more distant places such as Venezuela, Cuba, Amsterdam and the Canary Islands. Economic growth at this time is best explained by these inter-urban and regional connections and flows. Modernisation was a more sporadic phenomenon; nationwide patterns of development only really came after the mid-twentieth century. In the early part of the 1800s 'Spain' remained something of an abstraction. Palma had a small but well-developed community of international traders that not only dealt in the import and export of goods, but also of ideas (Ringrose, 1996: 198).

The rural landscape of Mallorca to about 1860

The improvements to agriculture that we noted in the previous chapter continued well into the nineteenth century: some estates were broken up, new enclosures and strip field patterns formed, the limits of farming extended and the new tree-based economy expanded. However, there were certain environmental constraints that were to hold back productivity for at least another hundred years, and in many ways still do so today. Soils remained poor despite increases in fertiliser application, water shortages at times were acute, and drought predominated until more sophisticated irrigation systems could be introduced. The years 1807, 1811, 1821, 1823 and 1845 were particularly dry. Rainfall remained erratic and unpredictable, often with extremes of precipitation that were damaging to the soil cover and to growing crops. Until the introduction of new breeds of both crops and animals that were more tolerant of the extremes of the Mediterranean environment it was difficult to diversify the agricultural mix. In this respect these factors were not so very different from those affecting much of mainland Spain too (Gonzalez de Molina, 2001). This was especially true for cereals and milk production. Mallorca therefore continued to rely on its traditional land-uses, expanding or contracting them according to market forces. The landscape effects of this could be very variable, but many of the landscape features of the century remain with us today, although others have fallen into disuse.

In Chapter 8 it was shown that there was evidence of the slow emergence of smaller units of landholding with their attendant field systems from the early modern period onwards, but land reform continued to be a problem, with the old feudal patterns still strongly present, since land was held in mortmain by the nobility, which gave them enormous jurisdictional rights over as much as 50% of all Spain, while the Church controlled about 17% (Shubert, 1990: 57). The removal of feudalism begins with the abolition of seignories in 1811, entail in 1836, and by 1840 the *ancien régime* had been largely demolished. This did not mean the nobility could not own land, indeed quite the contrary; even by 1930 they owned 6% of the land, and there were still nearly 2000 titled families in Spain, many of which exist today. Clearly in Spain there was no *à la lantern* on the same scale as in revolutionary France. Pressure for land-reform, however, did not initially follow the pattern that might be expected, that is, redistribution down the social and economic scale towards the landless peasants. Instead Liberal governments in Spain saw publicly held land as a redeemable resource that could be used to raise national income. This process, known as *desamorticazion* in Spanish, *desamortització* in Catalan and disentailment in English, was first practised in the late eighteenth century, but was most evident in the 1830s when it was used to help pay for the Carlist Wars, and under the government of Mendizabel in the 1850s (Tortella, 1987). Land that belonged to the Church, the municipalities and the state was auctioned off, but with compensation, to the highest bidders, who usually turned out to be either members of the new liberal middle class or, worse still, the old aristocracy. By 1875 about 90% of all such lands had been sold. In many parts of Spain the new owners proved to be no more efficient and productive than the old order; for example, wheat yields in Spain were only 7. 6 tons per hectare by the end of the nineteenth century, whereas Britain's had risen to 25. 3. Much of the land involved was economically marginal, which the new owners sought to bring into cultivation. Therefore, it is probably true to say that the landless peasants almost certainly lost out in this process because they had often cultivated these marginal lands unofficially, and now lost any traditional rights to them.

In Mallorca the disentailment process was a little different. In the Peninsula about 22% of the area was sold off, whereas on the island the figure was less than one per cent, and it was largely concentrated in the municipalities of Es Pla. Between 1820 and 1865, 76 rural fincas and 72 urban ones were sold off, primarily to new owners in the small inland towns such as Petra and Algaida, usually in such small lots as to make them unable to support a family so that many still had to work as day labourers. The average size of lots was about 6. 5 ha; only in Campos did it rise to over 14. In Sa Pobla 8 fincas were sold off in as many as 80 lots, giving an average size of only 3. 7 ha (Grosske Fiol, 1978–9). Although Palma's bourgeoisie participated in this land grab, only about 30% of sales went to them. However, they were successful in obtaining much larger units of land than the villagers. Church and monastic properties were especially susceptible to sale. For example, in Felanitx the Sant Augustí monastery and its Church, which dated from the late seventeenth century, was auctioned in1844 for 139,000 *reales*. In this case it was not torn down; the Church bought the convent back in 1901. In Palma the enormous Gothic convent and church of Sant Domingo, (Fig. 9. 1) dating from the fourteenth century, which stood on the site now occupied by the Palacio March and the Parlament on Calle Conquistador, withstood the first attempt at disentailment in 1821, only to succumb to the demolishers' hammers and crowbars in 1837, despite a last minute change of heart by Mendizábel (Bestard, 2011: 147–150) (Fig. 9. 2). Such destruction robbed the historic townscape of one of its iconic landmarks but, as is shown below, it yielded very valuable land for future urban development.

Figure 9.1 Sant Dominigo's convent – extract from Garau's map of Palma, 1644.

Figure 9.2 Calle Conquistador in 1910. This street was built partly on the site of the medieval Sant Domingo Convent. It was begun in 1295 and consisted of two cloisters, a chapter house, a massive Gothic church and a chapel. It was sold and demolished in 1837 as part of Mendizábal's policy of *desamortizació* and the site used for urban renewal including the construction of Calle Conquistador. Source: Muntaner.

The effect of disentailment in Mallorca, then, was much less than on the mainland; the nobility and the new middle and professional classes remained in a much more dominant position, imposing their backward views on land: the important thing was to own land, not necessarily to work it productively. By 1860 the aristocracy owned nearly 10% of the island's area – the Marques de Bellpuig alone owned 7635 ha (Bisson, 1969). The ennobled bourgeoisie owned another 5%, so that there were still more than 60,000 daily-paid land workers at that time, about 29% of the island's population (Rosselló Verger, 1981: 24).

However, the nature of the basic foodstuff, namely carbohydrates, continued to change. If the large landowners and their tenants carried on with the low yields of wheat and other bread grains, the smaller farmers turned increasingly to pulses. By the late 1820s these legumes accounted for 13% of agricultural output by value, occupying nearly 12,500 hectares by 1860. Meanwhile wheat production remained fairly constant, although as yields rose the area under the crop shrank. The most dramatic fall in the early to middle part of the century was in olive production, the area falling 40% between 1818 and 1860 (Manera, 2001: 113). This decline affected the upland areas more than the lowlands, of course; over 40% of the island's oil came from the 11 municipalities of the Tramuntana in 1818. This pattern of land-use change can be observed through the middle of the period. Between 1835 and 1860 cereal production grew by 47%, but legumes by 73%. Almond output doubled, carobs tripled and fig production increased five-fold. Output of olive oil, on the other hand, grew by only 37%. From the second half of nineteenth century the presence of cereals and olives in the Mallorcan countryside was to decline. By 1875 the tree crops accounted for 25% of agricultural output by value, rising to 40% in the 1920s (Manera, 2006: 66–67). The table below gives a snapshot of agricultural land-use in 1860.

Table 9.1 Land-use in 1860. Source: Barceló Pons 1962.

Type	Ha	%	Crop	Ha
Irrigated	6187	3.2	Cereals etc	4746
			Almonds	647
Dry	190526	52.6	Cereals etc	122779
			Almonds	5314
			Carobs	7610
			Figs	12798
			Olives	25949
			Vines	15543
Uncultivated	165074	45.6	Pine	10222
			Holm Oak	55035
			Monte Bajo	68466
			Wasteland	46787
Total Cultivated	197116			
Total Mallorca	361790			

1860–1920: fields, tree crops and settlements

Arboriculture and the landscape

The pressure for further land reform began to grow from the mid-nineteenth century. Mallorca's rural population was increasing, but country people were not able to support their families by working as irregular agricultural labourers; they needed land of their own, and there was only one source for that, the traditional large estates of the nobility and the smaller, but still large, lands associated with *possessions* – see Figure 9. 3. From about the 1860s the rural landscape of Mallorca was to be transformed by two forces. The first was the subdivision of the large estates (parcellisation), usually following the *establiments* system, to give the patchwork of very small holdings and fields of one or two *quarterades* that by and large have persisted into the present era. This did not mean that the *possessió* disappeared; by the late nineteenth century that were still more than a thousand of them, 40% larger than 100 ha (Valero i Martí, 2011: 78). The second was the introduction of new tree crops that clothed so many of these new shapes on the ground, giving a new appearance to much of the island. Two other landscape factors were to stem from these changes. One was the construction of a network of rural lanes and trackways (the *camis*) that were needed to give access to the new fields. The other was the gradual dispersion of population out of the old nucleations into farms closer to new fields, making for a more efficient system, reducing travelling time from home to fields. Although in many ways all these

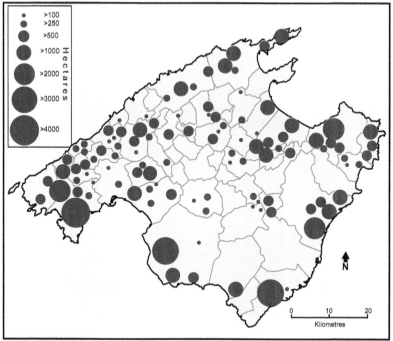

Figure 9.3 Map of large landed estates (*possessions*) in 1870. Source: adapted from J. Bisson.

were continuations of changes that had their origins in the initial break-up of the medieval estates, this late-nineteenth century process was probably the most dramatic of all the movements that have fashioned the rural cultural landscape of the island.

In a detailed study of three municipalities undertaken in the 1970s, the French geographer Jean Bisson was able to show that although there was some variation in these processes in terms of the types of land and its original ownership within Mallorca, there was a remarkable consistency in the levels of redistribution. For example, in Calvià in 1863, 89. 8% of the new holdings were less than 5. 0 ha. In Selva in 1860 this figure was 91. 2%, and in Vilafranca in1870, 96%. Some large estates, of course, remained intact – in Vilafranca one such of over 1000 ha covered nearly 48% of the municipality in 1870, although by the 1960s only two estates covered more than 100 ha. There is no real evidence to suggest that the land distributed to the small *pagès* was not of the better quality in terms of soil types and access to water supplies, but observation suggests that the more accidented areas and those around the existing nucleations were broken up first. While it would be wrong to describe this as the end of the latifundia system, it was the beginning of a rapid transformation of the Mallorcan countryside (Bisson, 1977: 107–134).

The land reforms, slow and patchy as they were, enabled the commercialisation of agriculture to increase. The countryside began to seem less and less one of neglect or merely of subsistence. The idea of the 'cash crop' began to take a more prominent position, especially after mid-century. In addition to the tree crops we saw emerging before 1820 – almonds, carobs and figs – a number of other crops have been studied in some detail: the vine in the central regions of the island, citrus fruits in the Sóller valley and later, potatoes on the flatlands around Sa Pobla and rice in the reclaimed wetlands of Albufera in the north. Tree crops were widespread throughout the island, though over time areal specialisations began to occur such as growing apricots in Porreres.

Almonds (*prunus dulcis*) formed part of the polycultural landscape that is now considered archetypal of so many parts of the Mediterranean, with cereals being grown in rotation under the trees. In the days before tractors, when the plough was mule-drawn, this combination proved profitable on the less fertile soils. By 1810–20 Mallorca produced between 1000 and 1500 metric tons each year. A century later the island's output reached over 7800 metric tons, representing more than 30% of all goods by value. Although forming part of the landscape of the fields of the smaller farmers, almond trees were also grown on the larger seigniorial estates, partly because the labour input was relatively small, being confined to harvesting and tree lopping. However, until new machinery was introduced, shelling the nuts was costly – one reason why most of it was undertaken by families, so that real labour costs were rarely taken into account. Like olive oil, almond production was tied to the soap-making industry as well as food processing. Later the discarded shells were to become an important fuel for the tile industry, especially around Vilafranca de Bonany. The growing of almond trees became one of the great staples of the Mallorcan rural economy from the mid-1800s onwards, expanding and contracting according to demand, but nonetheless a fairly constant element in the countryside. In later chapters we will show their considerable expansion in the middle years of the next century. By the 1930s there were nearly 50,000 ha growing almonds. For the tourist the trees are an attraction at blossom time in February.

Carobs (*ceratonia siliquia*) and figs (ficus carica) were the other major members of the Mediterranean polycultural tree economy, their growth expanding from the late eighteenth century. Like almonds, they were well suited to Mallorca's alkaline soils. The dried fruits of the carob could be ground up and used as a substitute for flour, eking out poor diets in times of poor wheat harvest and high prices. A gum produced from its seed was widely used in the food industry. Trees were planted at low density and took about five years to grow from budded stocks, much longer in times of low rainfall. The crude harvesting with nets and poles accounted for about a third of total costs. By 1860 carobs made up 4% by value of Mallorca's agriculture.

Figs are amongst the oldest domesticated plants, probably predating grains and legumes, so that they were well known in Mallorca throughout the historical period. Their commercial production dates from the seventeenth century, but expanded considerably in the nineteenth when their fruits, instead of being used to supplement human diets, were fed to pigs, their high carbohydrate content, in the form of sugars, enabling these voracious animals to put on weight very rapidly. We were able to show in earlier chapters that although the pig has a long history in Mallorca, it was a minor farm animal compared to sheep and goats, except on the domestic scale. Traditionally pigs had been herded with the holm-oaks in the Tramuntana and fed on acorns (pannage) though on nothing like the scale on the savannah of Estremadura (Grove and Rackham, 2001: 204–7); feeding them on figs enabled a more commercial approach. It was their increase in the eighteenth and nineteenth centuries that led to country families turning to meat eating. Until that date diets had been essentially vegetarian. Eventually this was to lead to specific meat processing industries such as the local *sobrassada*, but not the Serrano hams, because of the dampness of the island's climate. By the 1870s figs in another form had begun to replace oranges in the Sóller district, mostly dried and packed for export to France, a trade valued at more than one million francs (Bidwell, 1876: 91).

Viticulture in Es Pla, 1875–1910

Earlier chapters have shown that the vine and winemaking had a long presence in the Mallorcan landscape, even under the so-called temperate Arabs. In the seventeenth and eighteenth centuries the area under vines increased, and wine began to emerge as an important product which, rather like olive oil and many other tree and bush crops, generated income to pay for the import of bread grains. On larger holdings the vine had to compete with wheat, both requiring similar soil and water conditions, but rarely in the1800s did it exceed 25% of an estate. For the small tenant farmer, his newly acquired strip holdings could be used flexibly to grow crops, including vines, which might fetch higher prices. A number of factors were important in the decision-making process, relative prices being the major determinant, but taxes another. During the Napoleonic Wars taxes on wine rose, and much land reverted to wheat, a more valuable food crop. From the mid-eighteenth century two main geographical areas accounted for three-quarters of island production: Binissalem in the raiguer and Felanitx in Es Pla, the former concentrating on table wines mostly for home consumption. In the zone around Felanitx the wines tended to be acidic, and were best used for distillation. Spirit-making, ranging from brandies to industrial alcohol, increased the value-added content, making them more suitable for export. By the 1840s Felanitx produced more than 615,000 litres of spirits annually from twenty distilleries. Initially this all had to be exported by road to Palma at considerable cost, but one which the higher value brandies could absorb. Local producers realised that if a nearer port could be used transport costs could be reduced. The most suitable outlet was Portocolom, a well-sheltered harbour to the east. However, this did not have a customs office, and the road from Felanitx to the port was in poor condition. Legislation and investment in both aspects in the 1850s made this new route for exports possible (Fig. 9. 4).

Figure 9.4 Portocolom became a commercial port in 1855 but the wharves, seen here in 1920, were constructed in the 1880s to accommodate the boom in wine exports from the Felanitx area to many parts of Europe, but especially to France, where phylloxera had ravished the vines of the Midi. The houses in the centre of the picture acted as warehouses and homes for the *négotiants*. By the end of the century the boom was over, and by the 1920s there were fewer than 70 houses in the port area. Today they command high prices as prestige residences. Source: Muntaner.

By 1891 the increase in the area under vines to a peak of over 30,000 ha can be accounted for by one major factor: phylloxera, which struck wine-producing areas in France in the 1870s. Thanks to a well-organised mercantile infrastructure, the island was able to react quickly to this new export opportunity, another example of Mallorca's more modern economic and trading outlook. For about twenty years the two major wine-producing areas of Mallorca underwent a landscape transformation as more and more land was put down to vines, not this time destined for the distilleries, but as wine *qua* wine. In 1885 Felanitx had 4000 ha under the vine; by 1891 this had risen to 30,000 ha. Wine production was only 182,000 litres in 1885, rising to nearly 15,000,000 litres by 1891. Thanks to the investment in transport infrastructure in Portocolom, a new export route was now open to the southern French ports of Marseilles, Sète and Toulon. This was a boom period for the smaller farmers as well as the merchants, but like all booms it was not to last. In 1891 phylloxera struck Mallorca too, and by 1893 wine production had plummeted to 1. 2 million litres. New strains of resistant vine were introduced into the Midi from America, and Felanitx's prosperity was largely over, victim of an overdependence on a monoculture (Manera, 2001: 172). There are relict features of this heyday in the town if not the countryside. On Portocolom's waterfront the quays and the storage areas for export have now been converted to desirable homes. However, there appears to be some evidence that phylloxera

Figure 9.5 The Felanitx bodega was opened in the 1920s when trade in wine was declining. This co-operative, which at its height had 500 members, was closed in the early 1990s. It remains as a relict feature in the urban landscape.

did not ravage the area under vines as much as was previously thought (Pastor Sureda, 2000). The actual area in the centre of the island in 1885 and 1895 is remarkably similar, at between 18 and 19 thousand hectares. In other words, the area simply fell back from its peak to its pre-phylloxera levels. Reductions in the export of wine were quickly compensated for by rising home consumption, although at much lower prices, and many areas of Es Pla converted their vine lands back to almonds, figs and cereals, which by about 1910 were becoming the dominant landscape elements. When production rose again after the First World War, local wine growers established a *celler cooperativa*, opening a huge bodega on the outskirts of Felanitx, today an empty relic (Fig. 9. 5).

One other boom crop of the nineteenth century was the growing of oranges in the Sóller valley beyond the Tramuntana. As with vines and some tree crops, its success was partly attributable to the flexibility given by the small size of holdings in a valley once dominated by historically large fincas, plus access to the only significant port on this mountainous coast, rather like Felanitx's access to Portocolom. In addition, it was a response to a growing demand in urban France, and largely developed because of strong trading links between Mallorcan émigrés in France and the Sóller region. Although sour oranges had been introduced by the Arabs many centuries before, Mallorca had developed sweet, dessert varieties that appealed to sophisticated palates. Exports to the Spanish mainland began in the late eighteenth century, but the real growth came later, from the 1840s when France's Mediterranean ports were being connected to the national rail network, enabling distribution throughout the country but especially to Paris. The British Consul, Charles Bidwell, remarked that $200,000 of orange exports passed through the port of Sóller in the 1870s (Bidwell, 1876: 91) but this value was already well below the annual peak production of almost 400,000 kg, which had been reached in the early 1840s. By 1926 Henry Shelley could report that the port of Sóller was almost deserted (Shelley, 1926: 167).

Changes in the agricultural landscape from the 1870s

The last quarter of the nineteenth century and the early part of last century was the period in which major changes to the face of the countryside were to take place, and it was linked to demographic changes. From the 1860s through to the late 1870s considerable population increases took place in the Pla and the *part forana*, but Palma showed negative changes. Much of this can be explained by the scourge of diseases such as cholera, German measles and diphtheria, yellow fever and smallpox, all much more damaging in urban environments; the more dispersed population of the part forana was much less affected (Barceló Pons, 1970). In the decade following the 1877 census the natural increases (birth minus deaths) in the rural areas were higher than those of Palma. This led to two things: population migration from Es Pla to Ciutat, but not sufficiently to reduce the demographic pressure on land resources, which in turn led to a crisis of land hunger. Overseas emigration was also a partial solution to this problem; between 1887 and 1900 over 20,000 people migrated overseas with an additional 11,000 in the first decade of the twentieth century. The land that might be made available for redistribution to lower down the rural social scale was in fact limited to the central areas of the island, together with parts of the marinas in the South. The lands in the Tramuntana and the Llevant hills were either unsuitable for cultivation or remained in the hands of the remnants of a reactionary aristocracy/nobility. The coastal regions were still dominated by uncleared garriga or were plagued by malaria, making them unattractive. The Albufera was similarly 'out of bounds' at that time (see below). The potential for land reform via further parcellisation lay primarily with Es Pla. The social divisions in the countryside had become extreme between the senyors, pagesos and the day labourers, whose employment was never guaranteed, with both men and women living on very low wages and subsisting on a largely meat-free diet. In this area many new small holdings were provided by the break up of the large estates, often reluctantly, for a new breed of small farmers. However, the units provided were often far too small. The quarterades were rarely able to support the new class of proprietors 'de petites parcel. les'. Although the gradual entry of the Mallorcan economy into the new world order of capitalist production and trade, especially for exports, helped create this new class, small producers were often vulnerable to fluctuations in market conditions and prices, as we have seen in the case of viticulture and phylloxera. It was often a case of 'boom and bust'; those small farmers who were flexible enough in their crops and production methods survived, those of a more conservative nature did not. Without the economies of scale these new landowners were stranded, finally succumbing when tourism took off in the 1950s. The landscape inevitably changed with them.

The agricultural landscape of Mallorca by the late nineteenth century showed many changes from that of the end of the early modern period. Although cereals and olive production still dominated the plains and the mountain areas respectively, the scenery was now considerably modified by the rapid expansion of the tree crops of almonds, figs and carobs grown on much smaller fields, the product of the break-up of many of the large estates. The vine had spread rapidly over the central areas of the island, and niche cash crops such as oranges and other citrus fruits had taken advantage of changing economic and infrastructural conditions where environmental conditions allowed. The central problem remained that identified in earlier chapters, namely that Mallorca always found it difficult to meet its requirements for basic foodstuffs such as wheat for bread, particularly as population numbers rose throughout the nineteenth century, from 204,000 in 1857 to 250,000 in 1887, to 270,000 by 1920. By its end, many of the rural areas were simply not able to make a living from the land, and population numbers did not increase in the last two decades of the century. Despite the fact that Palma contained about a quarter of the island's people, there was no spread of large scale urbanisation and industrialisation as there had been further north in Europe to absorb these numbers. The only alternative was emigration, the tearing up of centuries-old roots to search for new lives in the remnants of the Spanish Empire, and the Spanish speaking parts of Latin America. Mallorca had often lost population abroad in the past, mostly to the Peninsula, France and Algeria, but the loss from the 1890s to the 1930s was catastrophic, especially after the rising optimism of the 1850s to '80s. Pere Salvà has calculated that between 1878 and 1910 Mallorca had a negative population balance of 41,453, mostly from the *part forana* (Salvà Tomàs, 1992: 406–7). The day-labourers and the smaller farmers from the central areas were especially hard hit.

Population dispersion and agricultural colonies
We have already drawn attention to the increasing dispersion of the rural population following land reforms from the old pueblos, many of which were founded under the 1300 *ordinacions*, particularly in Es Pla. However, there were marked place-to-place variations, often dependent on the degree of land reform, the tenacity of the large landowners in resisting change, the response of the new crops to soil and water conditions, and general attitudes to change. In a detailed study of the south-east of the island Rosselló Verger was able to show that by the 1880s little dispersion had taken place in Campos (resistant large landowners), much more had taken place in Felanitx (prospering viticulture) while Santanyí showed signs of population spreading, thanks to the new tree crops on recently created smallholdings. By 1900 only 6.7% of Campos's population was dispersed into truly rural areas, while in Santanyí and Felanitx the figures were 12.6% and 24.0% respectively (Rosselló Verger, 1964: 152–56). It was in many of these new dispersions of people that settlement occurred in small, often primitive houses,

which were to be vacated in the 1950s and'60s, producing what Rosselló Verger called *una comarca decadència* (an abandoned region).

Meanwhile in the Tramuntana a somewhat different pattern began to emerge. In the truly mountainous regions to the north, on the large estates, their dependent populations remained concentrated. Where there was a terraced landscape, as in areas further south, then more dispersion took place, but often to quite large but new isolated farms. The frontier of settlement was also pushed into the forested areas as the woodland resources were exploited more vigorously for charcoal, lime and turpentine (see below). Where municipalities in the Tramuntana had well-developed towns such as Sóller, Andratx and Pollença, more dispersion accompanied their expansion. To the south, places such as Estellencs and Esporles simply increased population under the influence of regional expansion centred on Palma (Ferrer Flórez, 1974: 91–110).

One solution to regenerate the countryside that was tried, and which was to have an enduring effect on parts of the settlement pattern of Mallorca, was the establishment of a small number of agricultural colonies. Various laws passed from the 1850s and into the 1900s allowed large landowners to establish new agricultural settlements, providing they were some distance from existing villages. They were usually carved out of the large estates. One example is Colonia St Jordi which was developed on the Marquès de Palmer's lands. A second is the present-day town of Porto Cristo, originally founded in 1888 as the colony of Carme and made out of part of La Marineta, owned by the Marquès de Reguer, 14 ha in size and 12 km from Manacor (Fig. 9. 6). This was an area dominated by garriga, covered with pines and growing little of agricultural value. Founded in the 1880s and divided into small parcels, as was the case in many such colonies, the newer tree crops and vines were to form the economic base. Houses, churches and schools were often incorporated into the plans (Pastor Sureda1977–8: 175–7; Duran, 1980; Oliver Costa, 2003). Few colònies were successful agriculturally, and many were hard hit by phylloxera after 1891. Some were to form the nucleus of larger coastal settlements in locations traditionally avoided, later to become important centres for tourism such as Porto Cristo, which blossomed after the First World War (Duran, 1980). Their economic and geographical remoteness, such as that of Colonia St Pere in the north, often contributed to their failure. Those who gained most were often the original landlords who benefited from tax concessions. One poignant example was Pobla Nou in Alcúdia, dating from 1876. It was created by Henry Robert Waring, a member of the English Company draining the Albufera (see below). Its 200 ha grew figs, mulberries and poplars. By 1881 its population reached 93. It had a school and a church constructed thanks to Louis la Trobe Bateman, son of J. H. Bateman, who had converted to Catholicism. In 1955 the English visitor Charles Elwell tried to find Gata Moix or San Luis, as it was also known, but discovered only 'a few fireplaces and the ragged stump of a chimney', virtually all that remained of this philanthropic English venture (Elwell, 1995: 37).

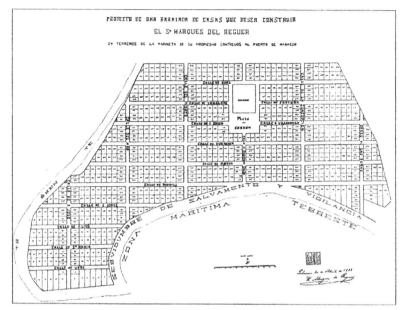

Figure 9.6 Plan of Colonia del Carme, 1888. This agricultural colony was built to accommodate farm workers and later fishermen at the old Port de Manacor, which, like most coastal areas, was largely devoid of settlement thanks to the fear of attacks by corsairs. Note the grid plan familiar in Spanish villages and towns. Today it is Portocristo, a seaside resort since the 1920s popular with British and French tourists. Source: Duran, 1980.

Woodland in the nineteenth century

In the previous chapter we saw that the process of disentailment transferred from communal use to private use some of the lands owned by the villages, the church and the state, using legislation passed in far away Madrid. This same process was also carried out in the wooded areas of Mallorca, especially in the communal lands of the villages there (Aizpurua and Galilea, 2000). Such areas were often perceived by outsiders as 'wasteland' but they were in fact a quite productive part of the rural economy, supporting fuel gathering, grazing, hunting and timber production, complementing the more formal field-based agriculture. To a certain extent the loss of these lands to new private owners was to have some adverse effects on the upland settlements, privatisation forcing such populations into other activities, often away from the land. The new owners, as in the agricultural sector, were usually either members of the new liberal middle class or the old aristocracy. Although they did exploit the timber resources, it was often in too rapacious a manner, without proper management, at the same time denying local people their traditional rights. Or at least that was legally the case, but as so often in such landscapes, new controls were difficult to enforce and many villagers carried on as before. This was especially true of the legislation passed in 1812. By the middle of the nineteenth century, however, the state began to take more interest in forest resources as part of 'national wealth', particularly for the navy and the railways, and drew up a catalogue of forest resources in 1859, the Clasificación General de Montes Públicos. In addition a wide variety of local markets for Mallorcan timber developed under the general impetus of economic growth. These included small-scale shipbuilders, charcoal makers, lime burners and turpentine distillers, as well as house builders. And from the middle of the century there was a serious danger of deforestation, so much so that guards had to be maintained in some areas to try to prevent it. As the century progressed, and well into the next, increasing pressure fell on woodland resources as an export market in charcoal and logwood developed. In 1884 Palma exported about 472,000 kg of charcoal, 400,000 kg of *curteja curtiente* (short timbers) and 1.15 m kg of logwood. Andratx sent out 375,000 kg of charcoal and the northern port of Alcúdia 490,000 kg. As the island and Spain had few coal resources, charcoal was a major fuel, particularly domestically. By this time the traditional source of wood for charcoal-making, namely underwood and branchwood, was proving insufficient, and rapacious charcoal-makers began to attack the trunks and timber of the forests themselves. Contrary to Oliver Rackham's usual thesis, is this one of the few irrational economic acts by woodland-users, that is, destroying the very resources upon which they depended? Rackham argues that many visitors to Mallorca commented on the bare hillsides and the loss of forest, and while it is true that certain areas were over-felled for charcoal production, industrial and constructional use and some shipbuilding, much of the island had never been heavily wooded. Indeed, by the late nineteenth century, Mallorca probably had more trees (if not continuous woodland) than at any time since the pre-Roman era, even if many of the trees were cultivated crops (figs, etc.) and regenerating Aleppo pines within the garriga (Grove and Rackham, 2001: 168–9 and 214; Berbiela Mingot, 2010). The round, flat sites surrounded by a low wall of stones that were used to make the airtight mounds of smouldering wood (sitges) are readily found in the woods of Serra Tramuntana today, but in the late nineteenth and early twentieth centuries many of the pinewoods of the coastal lowlands were also severely depleted, often by the *fexiners* (pine foresters) deputed to manage them. For these two types of area – mountain and coast – here was an example of quite rapid landscape transformation, much of it still visible today despite considerable regrowth over the last fifty years. Nonetheless, there remains a debate as to the relative contributions of over-exploitation and climate change as a cause of the reduction in forest cover, especially in the Tramuntana.

Albufera: a wetland landscape

The notion of a wetland landscape in the Mediterranean seems at first sight something of a contradiction in terms, but in fact such areas are quite widespread in lowland zones, particularly where rivers (even if intermittent) decant into the sea, perhaps forming deltas, and where rising sea levels in the post-glacial period have enabled offshore sand barriers to develop so that marshlands form behind. A more detailed explanation was given in Chapter 3. Such areas have offered a variety of opportunities to society over the last 2000 years: fishing, hunting of wildfowl, reed gathering, etc., and where dry sites existed, perhaps as islands, places for pasture and settlement. One such wetland in northern Mallorca is the Albufera, a Spanish term to describe such geographical regions. Despite the efforts of the tourist authorities to improve access to it, the Albufera remains largely unknown to most visitors, and yet its landscape has a fascinating history. Throughout Europe such landscapes were a product of a combination of varying degrees of entrepreneurialism, engineering, political will and environmental constraints (Cosgrove, 1990).

By the eighteenth century this landscape remained much as it had been for the previous millennium, but clearly the emerging, more scientific view of agriculture saw its potential. If technologies could be harnessed to lower the water table to reveal the rich soil resources fed by a more constant water supply than most parts of the island, then the area might be reclaimed. Small-scale reclamation had taken place around the edges of the wetlands and farming plots (*marjales*) (see Duran Jaume, 2003: 88–9) established in private hands, but the core area passed into the Royal estate. Throughout Europe from the 1600s onwards Dutch engineers had mastered the art of land drainage (for example, see: Lambert, 1971 and Ciriacono, 2006). Would it be possible to utilise such skills in Mallorca? As early as 1799 a plan had been put before the Sociedad Econòmica Mallorquina de Amigos del Pais for draining the Albufera, but it was not until the 1850s that progress was made. In November 1852 an enormous inundation known as 'plena d'en gelat' flooded a large area up to Sa Pobla and Muro, after which the engineer Antonio López was commissioned to draw up plans to drain a large proportion of the Albufera. His scheme included the canals En Ferragut, S'Ullastrar, En Molines and En Conrado, although these were never completed by him. The principal challenge, however, was to control the drainage of the two main rivers across the area, the torrents of Muro and Sant Miguel. This would require the construction of a 'grand canal'.

In the next decade the narrative begins to have particular interest for English readers. Perhaps the best account in English is given by Charles Elwell, mentioned above. [1] His researches reveal that in 1863 access to the Albufera was granted to the Majorca Land Company of London by royal decree. The directors were the distinguished English engineer John Frederick LaTrobe Bateman and a Colonel Hope of the Iberian Irrigation Company, who had probably heard of the potential of the Mallorcan wetlands whilst working in the Peninsula. Work began in 1863, using 1500 labourers drawn from all over the island for what was, to them, fabulous wages, small compensation for the atrocious conditions and rampant malaria. Pumping engines, bridges and roadways were built, and the projected Canal Grande – 50 m wide and 3. 0 m deep – was constructed, protected at its seaward end by two massive breakwaters (Figure 9. 7). To encircle the Albufera the 72 km Canal Riego was built. Altogether over 2000 ha of land were reclaimed, the freehold being granted to the English Company with Bateman Senior as the sole owner.

In order to recoup the considerable investment in civil engineering, the new lands would need to grow productive crops, but by 1877 only 343 ha were being cultivated by 483 tenants. Waring, one of the engineers on the project, estimated that as many as 1200 tenants would be required. Local farmers were only interested in low value crops like oats. Cotton was tried unsuccessfully; hemp and beans were more successful. Two other more serious problems were encountered: the high

Figure 9.7 The Gran Canal in the Albufera – the principal drainage channel in this wetland. Attempts were made to drain this area from the early modern period but the greatest works were undertaken in the 19th century by the Majorca Land Company, led by the British engineers Bateman and Waring. Source: Estop.

levels of salt water intrusion that prevented many crops from being established, and the rapid invasion of reeds, which flourished on the newly won lands. Frederick Bateman died in 1889 leaving his lands to his son, but by then the project was failing and the water table was rising again. The lands were sold to a Palma nobleman, Joaquim Gual de Torrella, who set up a new company, *Agrícola-Industrial Balear*, which used advisors from Valencia to try to establish rice cultivation.

Rice introduced yet another new element into the Mallorcan rural landscape: another attempt at diversifying agricultural production. Superficially the environmental condition seemed right, and if yields could be high enough the price structure might support this new crop. Mallorca, like much of the Catalan lands, was a rice consumer. The new owner had trials of rice-growing to try to ascertain likely yields. These showed that somewhere between 1600 and 2080 kg per hectare could be produced without fertilisers, a yield that was higher than those of the Ebro delta and Valencia. It was estimated that about 58% of the land reclaimed by the English Company was suitable for rice. On this basis Torrella went ahead and farmed the land on the familiar latifundia system, worked by day labourers. However, this proved no more successful than earlier ventures. The final blow came in 1906 when a huge flood led to its collapse.

From then on, Torrella's solution was to change the scale of production to a minifundia system with much of the land converted to small holdings (*finques petites*) for the people of Sa Pobla. The land was divided into small parcels (*velas*) between sweet water canals irrigating the fields and foul water canals taking away the pollution. These *velas* were very small, usually less than a tenth of a hectare in size. Irrigation was carefully controlled by a local *regidor* (Duran Jaume, 2003). By the 1930s the area under rice had fallen to less than 100 ha (it had been 260 ha in 1906) and although small quantities of rice are still produced on the *marjals*, commercial production did not last beyond the 1970s.

One last fling to try to raise some economic value from the Albufera came in the early twentieth century when a paper manufacturing plant was built, following Waring's original suggestion (Waring, 1877–8) (Fig. 9. 8). Established in 1922, it was based on the reedbeds as its raw material source, adding a factory and its chimney to the local scene, parts of which are still visible. In 1926 Torrella transferred ownership of his Albufera estate to Celulosa Hispanica SA and production continued until 1966.

Despite the efforts of Bateman and the English Company and those of Torrella and the agricultural workers of Sa Pobla, Muro and Alcúdia, the Albufera gradually through the last century began to revert to its more natural landscape of hidden waterways, reedbeds and bird and fish life; inevitably outbreaks of malaria continued, restricting any human activity to the margins. It was not until the rise of the environmental movement in the mid-1900s that the Albufera's ecological importance, and its potential for controlled tourist access, began once again to change the value that society put on its landscape.

Figure 9.8 The hydraulic pumping station at Albufera. The buildings behind formed the site of the later paper factory. Source: Muntaner.

The Archduke and the landscape

The Albufera we see today was a product of the skills of English and Mallorcan engineers seeking to transform a natural feature into productive land by reshaping and controlling the drainage system. This kind of large-scale interference with nature was not to be repeated until the coming of the mass tourist industry in the late twentieth century. However, landscape modification for *aesthetic* reasons has a longer history, best seen in Britain, for example, in the eighteenth century landscape garden movement led by Capability Brown, Humphrey Repton and William Kent. Mallorca had its larger gardens, such as Alfàbia and Raixa, but they were not on the scale of the British experience, and in any case tended to reflect earlier Arabic and Italian ideas (Valero i Marti, 2011: 84–101). They were gardens in a landscape rather than a gardened landscape; they were bounded by the property and not by the views from the house.

Archduke Luis Salvador d'Hapsburg-Lorena, fourth son and ninth child of the Grand Duke of Tuscany, Leopold II – a descendant of the Austro-Hungarian emperors, and Maria Antonietta of the Two Sicilies – first came to Mallorca in 1867. He has become something of a legendary figure on the island, partly because of his alleged prowess with local women, but more importantly for us, as a major influence on part of the island's western landscape. [2] In scientific terms he is best remembered for his magisterial *Die Balearen*, published in Leipzig in 1882. An

149

apocryphal story tells how he gradually began to acquire land on the western side of the Tramuntana in order to save its trees from the ravages of the charcoal burners. He began with the Miramar estate in 1872, at whose heart could be found the remains of the language school of the great medieval Mallorcan figure Ramon Lull. This area descends almost vertically to the sea on one side and rises up through the folded Mesozoic limestones to mountains of the landward side. It is a heavily dissected area covered with holm-oaks and pines. Between these two extremities lies a relatively flat, intermediate area, the *replas,* 500 to 1000 metres wide at an altitude of about 400 metres. On this platform can be found most of the cultivated land, the main road between Valldemossa and Deià, and all the houses on the Miramar estate. A number of natural landscape elements appealed to the Archduke: the Foradada arch at the foot of the cliffs, the numerous springs and various caves. The second of these could be utilised for water features. The caves had been used for centuries, but were now perceived as having potential for fashionable 'grottoes'.

The Valldemossa municipality was sparsely populated at this time, with only about 1500 people, living mostly in the main village. It had gained a certain notoriety following George Sand and Chopin's uncomfortable visit to its Carthusian monastery in 1837 (Sand, 1855). Two large estates were at its core, covering nearly 70% of an area given over to olives and wheat, with extensive woodlands providing fuel for charcoal and lime making. For most it provided a poor living, leading to some emigration to Latin America and France. The estate itself consisted of twelve farms acquired in the thirty years after 1872. The Archduke's final holding totalled over 1600 hectares, about a quarter of the old Valldemossa/Deià area. Most of his land grew olives and carobs interspersed with large areas of *garriga,* but deriving income from the estate was not the principal motive. Rather, it was to exploit the area's heritage in order to create a Romantic landscape to be characterised by a respect for the past, to indulge his taste for exotic construction and to conserve nature – what Cañellas Serrano has described as creating a medieval landscape, *'grandiós i fantasmagòric fons de muntanyes amb torrents i salts d'àgio amb els que harmonitza millor un cel ennigulat'* (large and fantastic mountains with streams and waterfalls set against a lowering sky) (Cañellas Serrano, 1997: 118). There was also a ludic aspect to his views on landscape. He was, after all, only twenty-five when Miramar was created, not a man of maturity. So we find elements of Roman and Greek architecture and Arab irrigation jumbled together. On his serious side he wanted to compose a religious aspect to his land, largely Christian but also including reference to Islam, perhaps in deference to Lull's links with North Africa and the Middle East, but also to Mallorca's own Muslim inheritance. Most of the work was carried out by local labour and supervisors because the Archduke, like so many of his noble Mallorcan contemporaries, was an absentee landlord.

In the thirty-odd years that he was active in shaping his estates, he restored Miramar, rebuilt its patios and replanted its gardens using local species. The chapel, its cloisters and the *hostatgeria* (originally lodgings for visitors) were all restored with contemporary materials. The San Ferrandell possession further south was in ruins, but brought back into use. Son Marroig, one of his most dramatic works, was given an Italianate garden, an Arabic *aljub*, marble balustrades and an Ionic tower. The Mirador des Puig de Sa Moneda imitates the large mosque of Kairouan in Tunisia. Parts of the small chapel of Puig des Verger have an Arab cupola. On the religious side he tried to create a landscape of ground and buildings that reflected the original function of much of the area as a place of pilgrimage, so that one can find chapels, hermitages and other elements of religious geography.

But the Archduke was keen to ensure that most of what he designed and had built above the wild cliffs of Mallorca's north-west coast was accessible to others, originally to his many noble and well-off friends who came to visit him, often on his large yacht, the *Nixe*. Viewing points, miradors with stone benches, picnic shelters, grottoes, imitation defensive towers designed to look like the *atalayers* to warn of corsairs as at Son Mas, were all erected to provide interest and diversion, all linked by numerous footpaths through this vast area. Most of this landscape of Romance and contemplation reflected many of the high values of late nineteenth century *mitteleuropa* with respect to Nature. Fortunately much of it has been retained or restored, and is open to walkers today seeking a retreat from the madness that was to overtake so much of Mallorca's coastline sixty years later. [3]

Nineteenth century industry: workshops and factories

In the introduction to this chapter reference was made to the early industrial development of Mallorca as 'industry without a revolution'. Amongst the reasons why Mallorca was thought by many modern writers to be under-industrialised, at least in the first half of the nineteenth century, were two related factors: first, the perceptions of visitors to Mallorca, who saw it as *La Isla de la Calma*, the Forgotten Isles, as some kind of idyll; and secondly, by the time mass tourism began to take off in the late 1950s much of the landscape of industrialisation was destined to disappear. Indeed, the link between the two processes is obvious. Tourism did not want a landscape of factories and workshops, but instead sought to promote the early twentieth century visitors' perceptions of Mallorca as a 'paradise island' (Manera and Roca, 1995). In fact, early industrial development, when translated into landscape terms, was characterised by a sporadic rather that ubiquitous pattern, in which no real industrial towns or regions on the German, Belgian or British patterns emerged. However, although urbanisation, as a demographic process, may not have been on the same scale as in north-west Europe, the population of Mallorca did increase by 50% between 1797 and 1920. Palma's population increased by nearly 60% (see Table 9. 2), the number of towns with a population in excess of 5000 (excluding Palma) increased from six to thirteen, and their population increased by 63. 5%. Increases in the first half of this 'long' century were indeed faster than in the second half (see Table 9. 3). In spatial terms there were no coalfields or

Table 9.2 Population in Mallorca, 1797–1920.

	1797	1825	1860	1877	1887	1900	1910	1920
Mallorca population	136671	168110	209604	230396	249008	248259	257615	269763
Palma	31942	36617	53019	58224	60574	63937	67544	77818
No. of towns with > 5000 population (excl Palma)	6	6	8	10	-	12	11	13
Towns with > 5000 Total population	37805	49088	65072	80765	105322	92627	90086	103724
%age Palma	23.37	21.78	25.39	25.27	24.3	25.75	26.27	28.7
%age other towns >5000 population	27.66	29.19	31.13	35.05	43.3	37.31	35.04	38.45
%age population in all towns inc. Palma	51.03	50.97	56.52	60.32	66.6	63.06	67.15	67.29

Source: Adapted from Manera (2001).

Table 9.3 Palma's share of population 1797–1920.

	%age increase 1797-1920	%age increase 1797-1860	%age increase 1877-1920
Mallorca	49.34	34.8	14.61
Palma	58.95	39.75	25.18
Other towns with >5000 pop.	63.55	41.9	22.13

Source: Adapted from Manera (2001).

transport nodes to act as magnets for industry, on the basis of the classic least cost location theory, so that geographical concentration was initially in Palma with its port functions where most new forms of manufacturing first flourished, typical of a primate city. Later there was some dispersion to other centres lower down the urban hierarchy, and some totally new industries grew up in such locations, such as the artificial pearl industry in Manacor (Sansó Barceló, 2009). This last activity also serves to illustrate the role of inward investment in the development of manufacturing, since one of the characteristics absent from the industrial scene was the relative lack of local invention and innovation in Mallorca, together with a paucity of capital for large-scale industrialisation.

For much of the island's history manufacturing meant cottage or small work-shop industry as, for example, in the textile industry examined in the previous chapter. But this industry was to herald the introduction of a factory system of huge importance in developing something approaching a true industrial landscape. Nonetheless, such a landscape in the urban milieu of Palma was for most of the nineteenth and early twentieth centuries made up of larger-scale premises and small-scale craft industries working cheek by jowl. Indeed, the characteristic pattern of production in Mallorca was one of the horizontal integration of a large number of small enterprises, each making a quite limited range of items, brought together by a separate group of entrepreneurs for final assembly and onward sales. This type of spatio-economic organisation is still prevalent in Mallorca today in, for example, the woodworking and furniture industries of Manacor (Sansó Barceló, 2011).

Tables 9. 4 and 9. 5 below illustrate the general economic structure of all the Balearic islands over the nineteenth century (Manera and Petrus Bey, 1991:14–16).

Table 9.4 Evolution of active population by sector in Balears (%).

	Primary	Secondary	Tertiary
1787	53	15	32
1860	68	13	19
1900	70	15	15

Source: Adapted from Manera (2001).

Table 9.5 Distribution of Industry by type (to nearest %).

	1799	1856*	1900*
Food	---	53	35
Textiles	43	16	22
Distilling	23	---	---
Metal working	23	<1	6
Leather, shoes etc	3	8	4
Pottery, glass	3	7	5
Woodworking	2	<1	5
Chemicals	---	14	8
Paper	---	<1	2
Construction	---	---	---
Total	100	100	100

*figures for all Balears; remainder for Mallorca
Source: adapted from Manera and Petrus Bey (1991).

The first shows that after a century of industrialisation the island economy was still dominated by its primary sector, especially by agriculture. The second shows the early and continuing importance of textiles and the later growth of a broad range of manufacturing activities. Before the 1860s manufacturing was characterised by small-scale workshops. The average number of employees in food processing was less then 5. 0, in textiles only 6. 0 and in metal working 3. 0 to 4. 0; only in the leather and footwear industries did it rise above 10. 0 (Penya Barceló, 1991: 59). The key to the establishment of a factory system in the landscape was the cotton industry, developed from the 1840s when the first steam-powered cotton mill was opened by Villalonga and Co. in Carrer de Bonaire in Palma in 1847. This was powered by a 45 hp steam engine driving 4336 spindles and employing over 200 workers (Roca Avellà, 2006: 12). However, early textile industrial development was distributed in a very wide variety of locations across the small towns in the Mallorcan countryside. This was especially true of the handloom weaving of wool, hemp, linen, silk and cotton, which remained a cottage industry through much of the nineteenth century. Quite a strong symbiosis existed between Palma and the other towns in this respect. In the rural areas, such small-scale manufacturing was an intimate and essential part of a diversifying dual economy between working the land and the home-based workshop.

By the late eighteenth century the textile industry was the principal manufacturing activity on the island, producing about 40% of the value of manufactured goods and employing about 60% of the secondary sector workforce but this amounted only to something like about 2000 workers altogether. By the beginning of the nineteenth century many workers had avoided the power of the guilds by

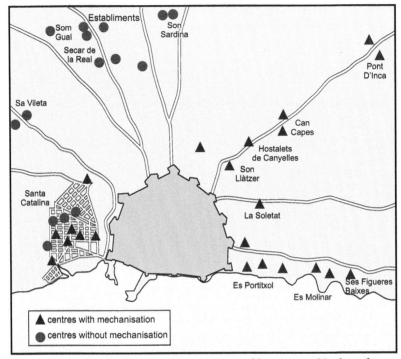

Figure 9.9 Textile mills in Palma in 1885. Powered by steam and built on factory system principles, they were located on the periphery of the city and in its new industrial suburbs such as Santa Catalina. Source: Manera and Petrus Bey.

working from home. This diverse and dispersed labour force was organised by a network of merchants that controlled the supply of raw materials, type of output, and was able to manage largely female workers essential to the spinning side of the industry, outside guild control. However, such a geographically and economically dispersed activity was difficult to mechanise and convert to a factory system. The Peninsula War with France saw Palma as one of the few ports less affected by the conflict, creating a new situation for the establishment of cotton factories away from Catalunya, the leading mainland region for textiles, but with which Mallorca had longstanding links. Once the first spinning mill was opened, then more soon developed, so that by the 1870s there was a score of them in Palma. The city, however, proved to be a difficult location, despite its apparent advantages. A military law of 1856 prevented factory building within 400 metres of the city walls, even though by that date the dangers of attack were surely much diminished. Similarly, local ordinances restricted boilers and steam engines – rightly thought to be prone to explosion – to the city's outer zones (Fig. 9. 9). A second mill was built in the fields near the present day Carrer dels Oms. Palma was still largely confined within its Renaissance walls, and by the mid-nineteenth century there were few left of the open spaces seen in the urban morphology of earlier generations. Santa Catalina and Soledad were two outer areas that were the location for new factories. In the latter, the Can Ribas company built one of the largest textile complexes, which specialised in cotton spinning initially, and later the manufacture of military uniforms. There were no water power sites in Palma as there had been in the upland areas for all the earlier textile mills. The new factories relied on imported coal for steam generation and were notable polluters in the city's confined spaces. Such constraints soon forced entrepreneurs to look elsewhere to develop their new industries. Nonetheless, by 1877, 49% of the secondary sector labour force was located in Palma, most of it still in the small workshops, so that despite the growth of the factory system it would be difficult to describe Palma has having an industrial townscape (Penya Barceló, 1991: 67).

One such inland town identified by Roca was Sóller, where the development of orange exporting (see above) had fostered a new entrepreneurial class locally and could provide capital for a range of industrial activities in what had been primarily an agricultural and market town. By concentrating on quality and profit margins, Sóller producers/manufacturers were able to develop sound small and medium-sized businesses, but based on *mechanised* workshops. By 1903 there were 304 looms, rising to 479 by 1920. All this stimulated the local economy, leading to the founding of the Bank of Sóller in 1889 and the opening of the railway to Palma in 1905 (Roca Avellà, 2006: 15).

Another activity that was originally concentrated in Palma and later developed in the *part forana* was the woodworking and furniture industry, a small-scale but well organised industry into which mechanisation and steam power were introduced relatively early. In the eighteenth century it was highly localised in Palma, with nearly half of all *fusters* located there, the remainder distributed throughout the island roughly on the basis of population. It was dominated by the guilds, with a hierarchical pattern of workshops each specialising in one product or one stage of production. Shipbuilding and fitting-out was concentrated in Palma, but cooperage was located in the wine centres of Binissalem and Felanitx. Furniture production was more widespread, as was the manufacture of carts and carriages. In the last third of the nineteenth century mechanisation brought about a shift in the industry, although some aspects of production, such as sawing, were more affected than others. It was not until the 1920s that the use of machinery was adopted throughout the industry, thanks to the spread of electricity.

A notable centre outside Palma was Manacor, today still describing itself as 'Furniture City'. In the early nineteenth century it developed a reputation for chair making, but industrial growth began with the setting up of larger-scale steam-powered workshops by Luis Lull Ferre and Juan Suñer Soler in the 1880s (Sansó Barceló, 2011: 61). In 1879 there were 17 wood workshops, 3 carriage builders, together with 18 upholsterers. By 1881 there were 71 workshops (employing 32% of Manacor's industrial workforce of 223) but still mostly in small workshops of less than five employees. Co-operative organisation of these after the 1920s led to further growth.

Railways, ports and roads: transport and the nineteenth-century landscape

Many visitors to Mallorca are surprised to find that there is a working railway system on an island so small, but given the points we have made about the industrialising economy in the second half of the nineteenth century, such surprise is perhaps unwarranted. Along with increased urbanisation, the development in many parts of Europe of the canalisation of rivers, of canals themselves, of metalled highways, the building of ports and docks and eventually of railways were integral parts of the modernising process, particularly with reference to industrial location, and distinctive landscape features in their own right.

Movements inland on the island were never good. There were few firm-surfaced roads before the second half of the nineteenth century, and goods and people were moved by slow-moving carts and trains of mules. Given the nature of the climate and soils, it is not surprising that early visitors complained of intolerable dust. The intense storm-like nature of so much of the rainfall meant that large sections of roads were often swept away, a not infrequent occurrence even today. Nor is it surprising to learn of limited mobility among the local population; visits to local market towns were infrequent, and to Palma very rare. The indefatigable Archduke Luis Salvador had a penchant for statistics, like so many in the nineteenth century, and observed that in 1870 Mallorca had 152. 5 km of road built linking up the island's 48 villages and towns, with another 18 km under construction and 92 more planned, all second and third class roads (Archduque Luis Salvador, 1962, trans. Sureda: 139–44). The Archduke thought them well built and maintained, but according to others a journey of 20 km in a day was the maximum tolerable. For the most part this network survived well into the 1950s, although by then roads were tarmacadamed and better drained. In addition there were nearly 600 km of local roads (*caminos vecinales*), the familiar rural *camis* of today, which are only now being upgraded. In the upland areas, and especially in the Tramuntana, there was what the Archduke described as an *infinidad de caminos de herradura* (bridleways) many of which were paved with hard limestone setts and often constructed in long steps, dating from the middle ages, possibly from the Muslim era, linking the more isolated settlements.

The major means of transporting bulk goods about the island before the coming of the railways was by sailing boat, and the coast was dotted with small harbours, though few actual coastal settlements. The capture of travellers and of goods purloined was a common and very real fear until after the age of the corsairs, despite minor fortifications to such places as Portocolom, Porto Petro and Andratx in the sixteenth and seventeenth centuries (Soler, 2004: 139–144). Soler has two maps of the large natural harbour of Portocolom from 1859 and 1885. No village is shown on the first, simply a customs post and the oratorio on the headland. On the later map, drawn at the height of the wine exporting trade, the construction of the mole and the quays and the beginnings of the barrio, the core of the modern town, may be clearly seen together with the *colonia* on the headland, a planned settlement for a fishing community established to offset local agricultural decline and unemployment. Only from the middle of the nineteenth century were ports and harbours properly surveyed and protective breakwaters built and quays laid out; again, another component of the modernising of Mallorca. Examples that benefited from these works included Andratx, noted for its soap and charcoal exports, and Sóller for oranges to France. In the modern landscape nearly all such places have become important holiday resorts, and former commercial ports are now dominated by yachting marinas.

The coming of the railways in the 1870s was seen as a means of lowering costs, increasing productivity and labour mobility, improving linkages between raw material sources – and in the case of Mallorca, its ports – and markets, with points of production (see Fig. 9. 10). Increasing efficiency was used to offset the costs of construction. Railways are amongst the largest of earth-moving and civil engineering projects, and their impact on landscapes has been profound. Cuttings, embankments and tunnels were quite new landscape features in their own right as well as altering pre-existing land in the process of their construction. However, in Mallorca the lie of the land and the disposition of the major settlements in the *part forana* meant that lines were, for the most part, built on relatively flat land and, with some notable exceptions such as the Palma to Sóller line, few of these earthworks were characteristic of the early railways in more northerly parts of Europe or in the

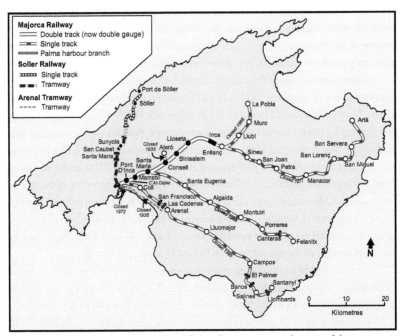

Figure 9.10 Map of railways. Railways formed an important element of the economic development of the island and its industrial landscape. Source: Barnabe and others.

Peninsula. By the 1870s locomotives were much more powerful than in the early days of rail development, when vast cuttings, enormous embankments and lengthy tunnels were needed to keep gradients at about the 1 in 500 mark.

Railways may have come relatively late to Spain, but in the case of Mallorca they were at least contemporaneous with developments in the Peninsula, especially in Catalunya – again evidence of the island's association with modernity elsewhere. The first proposals were for a line between Palma and Inca following the *raiguer*, an area of numerous towns and dense settlement likely to give adequate trade to warrant a railway. The early plans of the Belgian Paul Bouvy and the Catalan Gisbert brothers were disrupted by the 1868 revolution on the mainland. The first detailed study for the route was made in 1871 by Eusabio Estado, and the Majorca Railway Company was set up with the equivalent of £88,000 of local capital. A common and often erroneous view is that Spain's railways were all developed by overseas capital, but this line in Mallorca was 'probably the only Spanish provincial railway to be built without foreign capital' (Barnabe, 1993: 5). It was built, curiously, with a British 3 ft gauge, probably because nearly all the early *materiel* came from Great Britain. The stations and railway buildings were built to local designs, but with something of a provincial French flavour. The engines and rolling stock arrived by sea from Britain and the Palma–Inca line was opened in February 1875. Such was its success that other companies began to propose other routes to Llucmajor, and to Manacor via Sineu and Petra, this being the next line to be built, reaching Sineu by 1878, with a branch to Sa Pobla. Twenty years later a 42 km branch from Santa Maria to Felanitx was opened. In 1908 a line to Santanyí via Llucmajor was proposed, but because of topographical difficulties near Arenal, which required a high bridge and two viaducts, it was not operational until 1917.

Within about 40 years the basic network had been built, although the Inca–Manacor line was extended to Artà in 1921, when it reached its maximum length of 216. 5 km (Morley and Plant, 1963: 10) (Figure 9. 10). The traffic was two-way, bringing agricultural produce to Palma's growing population and its factories and workshops, and permitting some early commuting, especially on the Inca–Palma line, which accounted for 90% of passenger movements by 1920. In the opposite direction imported goods from Palma's harbour found their way to the inland towns and markets. One curious omission in the network was the failure to link Palma to the northerly port at Alcúdia. With the benefit of hindsight, another was the failure to link the coastal areas to the capital, a factor that could have been important in the development of the tourist trade in the mid-twentieth century.

One final and late addition to the system was the proposal in 1892 to build a line from the capital to Sóller, an undertaking that would require considerable civil engineering works as the line traversed the Sierra. The 1903 plan included a 2. 5 km tunnel at its peak, one of thirteen such structures that also included a long viaduct and many bridges and the laying out of many loops in the dramatic descent to Sóller (Fig. 9. 11). Originally steam-powered when it was opened in 1912, the

Figure 9.11 The Sóller Railway was built at the height of Sóller's economic importance that grew from the export of citrus fruits and cotton textiles. It was electrified in 1929. Today, it is a tourist attraction. Source: Wikipedia Commons.

line was electrified in 1929. Unlike other lines, this scenic route, together with the old-fashioned nature of the electric engines and the wooden rolling stock, has long appealed to tourists, especially as the line was extended from the start to the port of Sóller by an equally ramshackle electric tramway that rattles its way eccentrically through the streets of the town to the harbour.

The decline of the network began in the 1930s. By 1967 all lines were closed with the exception of those to Inca and Sóller. This left the landscape littered with the ghosts of the system in the form of the relict features of abandoned stations, crossing keepers' huts, goods yards, sheds and sidings, bridges and viaducts – even lengthy stretches of rusting lines that remain in situ. It was not until the revival of some of the routes in the late twentieth century that some of the distinctive architecture was either integrated into the new railway, such as Manacor's beautiful station, or new uses found for the old buildings, such as the art galleries at Artà and Sineu stations. In Chapter 11 the revival of the railways in Mallorca will be described as once relict features of the landscape are brought back into use.

Nineteenth-century Palma

The changing morphology of Palma after 1800 tended to reflect the social and economic pressures that were put upon it. At the beginning of the century the city's streets and blocks would have been recognisable to citizen of the 1700s; Garau's map of 1644 would still have been a useful A–Z. By the 1920s the *moderne* movement had brought about substantial changes in building styles, and the first generation of town planners had begun to propose major improvements to the road network. The liberal reforms of earlier in the century brought forth *desamortització*, the land sale phenomenon we saw at work in the countryside earlier in this chapter. In many ways perhaps the most noteworthy development came much later, at the beginning of the twentieth century, when the demolition of the Renaissance walls permitted the expansion of the city on and beyond their lines. The longstanding relationship between society and the built form of Palma that had existed through much of the medieval period and nearly all of the early modern period – that is, for about 400 years – was largely transformed by the effects of modernisation in the later parts of the nineteenth century and the first decade of the next.

The process of *desamortitzacio* released many new spaces within the old city, particularly for new civic buildings, often permitting redevelopment on a large scale. Six convents or monasteries were demolished, and many more suffered a drastic reduction in the area of land they commanded in the city, although most of their attached churches usually survived. For example, the church of Santa Margalida in Calle St Miguel, founded in 1238, became part of a military hospital.

A good example of religious space converted to public already mentioned was from the lands of Santa Domingo and St Francis de Paul just to the east of the *rambla* of the Born (Figs. 9. 1 and 9. 2). Demolition of this large block created an opportunity to remodel the hill slopes around the cathedral, with its stepped townscape, and the west side of Palau Reial with its porticoed flank (Lucena *et al.,* 1997:107). Many demolitions came later in the century as pressure on space increased. Consolaçion gave rise to Plaça Quadrado, opening up a network of narrow lanes and roads. Concepcion del Olivar eventually supplied land for the site of the new Olivar Market, which was not built, of course, until much later in the 1940s.

Two significant parts of the modern townscape were created in this century. The former river bed of the lower Riera (the rambla) had probably always been an informal paseo. In 1812 it was proposed that it become a new route cutting through the low town towards the port, but various delays resulted in its conversion to a more formal open space in the 1820s with subsequent decoration and embellishments completed in the 1870s (Fig. 9. 12). The second led to the creation of the Plaça Major, which was first defined in 1862 on an elevated site with a steep drop to the north-west in an area cleared of buildings originally used by the Inquisition. With the addition of land made available by the demolition of Santa Espiritus, it became first a fish market, and following further improvements in the twentieth century, *'el gran ejemplo de área creada para función publica utilaria'* ('a good example of an area created as public open space') (Lucena *et al.,* 1997: 106) (Fig. 9. 11).

By the 1860s the growing numbers and political importance of the new mercantile and professional middle class was demanding better accommodation, and

Figure 9.12 The Rambla in 19th century Palma: originally a section of the Riera river that ran through the city but was diverted around the new city walls in 1613. This photo of 1906 shows its recent improvement, a scene for the *paseo*, part of the theatrical townscape of Palma. Note the gas lighting dating from the 1860s. Source: Muntaner.

the area climbing up from the south end of the Born to the higher parts around La Cort was identified for offices, banks, shops and blocks of high grade apartments. From the new Plaça de la Reina, broad streets such as Conquistador and Calle de la Victoria were built (Fig. 9. 2). To the north of the Born, Calle Weyla was later pushed through to link it to Roma and the back end of the Plaça Major, including in its stretch such public buildings as the Teatro Principal (the modern Opera House) built earlier in the 1850s in the neoclassical style.

Meanwhile more piecemeal improvements to the internal network of streets was taking place by widening and straightening medieval thoroughfares to create, for example, Calle de Colom between Plaça Cort and the new Plaça Major. However, by the 1870s the emphasis in Palma's future design had shifted towards the need to create much more space by the demolition of the walls; from about the 1880s, *'abajo las murales'* became a popular cry!

In 1840 nearly 82% of Palma's population lived within the confines of its walls; by 1887 this had fallen to 69%. We have also noted that throughout its history the city contained a disproportionate share of the island's population of at least 25%; by 1930 this was to rise to 30%. These figures are significant for two reasons: even without creating more space through demolition, the population was moving inexorably beyond the walls, usually to unplanned urbanisations over which the Ajuntament had little control, and often following the decentralising industries identified earlier in this chapter. Secondly, the total population was

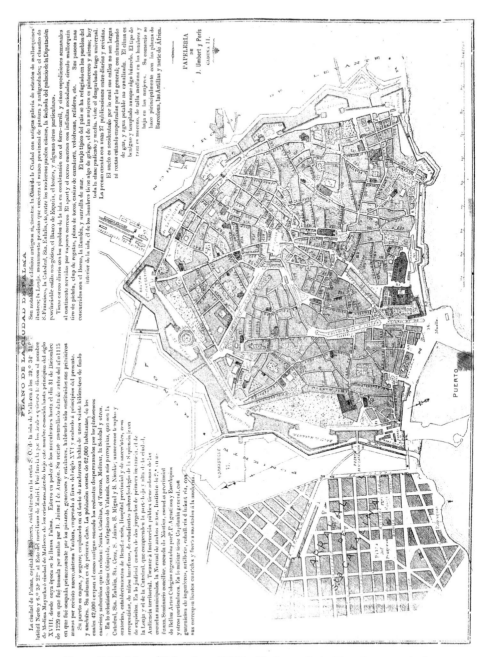

Figure 9.13 Map of Palma from 1888 by Umbert y Peris. Note the survival of the Muslim/medieval historic core and the 17th century walls, which would be demolished at the turn of the century. Source: Tous Meliá.

rapidly increasing in an absolute sense. There were many consequences of this latter point, but to many observers of nineteenth-century cities the most pressing concern was for the health of the citizenry, particularly of the prosperous middle classes, many of whom had no country mansions to retreat to at the time of plague and other infectious diseases. Palma was said to suffer from the miasmic effects of sewage and rubbish dumped in the narrow, unpaved streets, largely the result of the overcrowding of very old buildings without water supply; birth rates in the city rose from 0. 36% in 1840s to 2. 36% a decade later, and inward migration in the latter half of the century increased the city's population to more than 77,000 by 1920. In addition there was the fear amongst the better off that the low morals of the city's poor might lead to revolution, a common fear in so many European cities in this century. Palma needed more space into which to expand, and the only possibility was that which could be created by taking down the walls and the associated military works (see Fig. 9. 13).

The walls, however, were the property of the military, not the Ajuntament, and an agreement had to be struck that compensated the national government for their loss. The military gave the land to the city, but up to 40 ha of nearby land was asked for in exchange (Ladaria Benares, 1992: 54). The city engineer Eusabio Estada, the municipal architect Gaspar Bennàzar and the Commander of the Engineers drew up plans covering over 16,000 m² of land; demolition began in 1902 with the seafront walls between St Pere and Puerta del Moll and concluded with the demolition of part of the wall at Revellin de San Fernando twenty years later (Fig. 9. 14).

Figure 9.14 This photo shows that much of the city wall had been demolished by 1910, especially east of the Llotja. Large sections of the seawall and the *baluades* (bastions) remained at this time. Note the maritime activities on the quays and moles, including fishing nets and boatbuilding. Mule-drawn trams ran along the first *paseo maritima*. Source: Muntaner.

On either side of the encircling walls the land was intensively occupied by houses, factories and businesses; there were thirty-three different landowners involved. The plan not only involved new housing but new roads linking new suburbs to the old city and providing a ring road – the *avenidas* – around the *casc antic*. However, the process of redevelopment to create a series of new elements in Palma's urban landscape proved to be very long-winded. Of the 88,000 inhabitants of Palma in 1930, only 12,000 (14%) lived in *ensanche*, the new space created. Between 1903 and 1927, 368 building licences were granted for *fincas* (i. e. building plots for blocks of flats, shops, offices, etc. of two or more storeys; for example, in the *avenidas*), 469 for *casas* (detached houses) 232 for *cercar solares* (enclosures for building on) and 53 for warehouses and storage, but the majority of these came towards the end of the period (Ladaria Benares, 1992: 157). Although the plan had been to rehouse people from the overcrowded inner city, in fact, the principal beneficiaries were the middle classes and their businesses. As we shall see in the next chapter, the solution to this key problem – overcrowding and health – was not solved until the plans of the 1940s took effect.

Summary

Mallorca's nineteenth century was the first to witness such rapid change in the island's rural and urban landscapes since the Christian invasion in 1229. It was a period of modernisation, preparing the people for the even more dramatic developments that were to come in the first half of the twentieth century, themselves a forewarning of what was to come in the second half of that century. The period after 1830 saw the end of the Old Regime in social and economic terms and the emergence of a new bourgeoisie, principally with its geographical origins in Palma but with country estate, landholding and rentier aspirations. The landscape of the countryside witnessed new field patterns following on from the disentailment of the 1830s, saw the expansion of new tree-based crops, the adoption of a commercial vine and wine economy in parts of Es Pla, the taking in of new lands from the garriga, and the establishment of new agricultural colonies. New technologies transformed some landscapes, such as the Albufera marshes and the windmill-studded *marina* to the south. Population numbers increased, thanks to new economic opportunities. Some changed from a rural peasantry to an urban proletariat, but with many forced to seek a new life in the old empire – the beginnings of the Mallorcan diaspora – in the last two decades of the century and beyond. The traditional industries were considerably expanded by the introduction of a limited factory system, whilst retaining the small scale workshop patterns that persist today, giving Mallorca a quite different form of 'industrial revolution'. The city took on a new dimension, moving away from its original mercantile function to embrace new industries and businesses in its expanding suburban townscape. Its growth necessitated the traumatic destruction of its walls, an act symbolic of a new era, literally an opening out of the city into its proximate hinterland, rather than the old way of looking to the Mediterranean Sea and overseas as its outlet. Although little improvement was made to the road system, the introduction of the railways was to have a long-term impact on linking town and country, as well as introducing quite new landscape features. We have shown that all along Mallorca was always open to visitors, a veritable hub in the western Mediterranean, and while they now came for the same old and familiar commercial reasons, some were now arriving to take advantage of the island's paradisaical environment, giving a slow inkling of the tourist industry that was to develop once communications were improved with the introduction of steam ships (Buswell, 2011). The first guidebook for tourists appeared in 1891, Pere d'Alcantara Penya's *Guia manual de las Islas Baleares*. He drew attention to the sort of features that might attract – '*Sus paisajes, sus puntos de visitar son en gran manera pintorescos. Sus terrenos se cultivan con aquel cuidado y esmero que notamos en las jardines particulares. Los palacios del antigua nobleza continúen gran numero de objetos artísticos. . . y la historia. . . de las islas esta llena de interés, especialmente para los hijos de le gran Bretaña*' (Quoted by Serra i Busquets, 2001: 106) ('Its landscapes, its noteworthy places are very picturesque. The land is cultivated with the same care and attention that we observed in private gardens. The palaces of the old aristocracy contain many art objects . . . and the history of the islands is full of interest, especially for the sons of Great Britain').

Overseas travellers coming to Mallorca in the late nineteenth century were attracted by three things: the mild winter climate, the mountain landscapes to the north and west, and the historic sights of Palma. While the therapeutic effects of sea bathing were appreciated, the attraction of 'the beach' as we know it today was to come later. These visitors were to leave some valuable descriptions of the landscape that give an indication of the 'values' they ascribed to Mallorca's scenery. In the tradition of the early 'alpinists' visitors found Tramanuntana intimidating. Artists as well as entrepreneurs were beginning to seek out the pleasures of Santiago Rusiñol's Tranquil Isle, setting in train a new form of economic activity that was eventually to transform Mallorca's appearance almost beyond recognition.

10

A beggar's mantle fringed
with gold – Mallorca 1920–55

This is a difficult period in Mallorca's landscape history to introduce. In some ways it was a continuation of the last third of the 'long nineteenth century' (say, 1870 to 1920), a period of progressive modernisation that included the reform of agriculture, industrialisation and its attendant urbanisation and improvements in transport infrastructure. In another way it had about it something of the 'calm before the storm', the hurricane of mass tourism that was to hit Mallorca in the 1950s and which continues today. Perhaps this period can be described most aptly as one of transition. Unfortunately it does contain two world wars, but the effect of these on the island's landscape was minimal. The direct impact of the Spanish Civil War (1936–39) on the island's appearance was also limited, unlike in many mainland areas, with little physical damage by bombing and shelling. Its indirect effects on island psychology, however, were profound.

Three landscape themes are worth emphasising for this thirty-five year period: first, the continuing industrialisation of the economy. Despite the rising importance of industry as an employer, and its growing effects on the landscape, Mallorca remained largely an agricultural island, with changes also continuing in the countryside. Second is the growth of Palma beyond its Renaissance walls, which were demolished at the turn of the century, and the subsequent planning of new suburbs. Third is the emergence of the nascent tourism industry and the 'discovery' of the island by Spaniards and early overseas visitors with the development of seaside 'colonies', particularly for a new artistic and hedonistic coterie. James I of Scotland described one of his eastern counties, Fife, as *a beggar's mantle fringed with gold* – an apt description of Mallorca at this time, on the brink of its first boom of mass tourism when one of the oldest environmental resources of the island, the defining line that joins land and sea, was given a new economic and social meaning by being reinterpreted as 'beach'. Nonetheless, rural Mallorca remained a dominant component of the island's landscape.

The Mallorcan countryside of the 1920s and '30s
The countryside underwent change as agriculture became more productive, though scarcely mechanised. In the twenty years after 1910, thanks to the continuing importance of the trio of tree crops – figs, carobs and almonds – the total area under cultivation rose quite significantly by about 70,000 ha, its greatest increase since the early 1800s, with the area under the last two trees nearly doubling in the same

period (Manera, 2001: 250). This response in the rural landscape was largely due to the efforts of the newly invigorated small farming cadre, the island's growing food processing industry, and a lively export market. World prices for many of Mallorca's new agricultural products, both natural and processed, rose during the 1920s, encouraging intensity of output and experimentation. Many of the old problems associated with land reform remained, and some of the larger landowners took the opportunity to improve their estates, including drawing many of their workers into new nucleated settlements.

Thanks largely to increasing world demand and rising world prices in the 1930s, the trio of tree crops entered their most profitable phase. Almond tree planting doubled its area in the 1920s from 27,000 ha to nearly 50,000 ha. Similar levels of expansion took place with figs and carobs in the same decade (Casanovas, 2005: 50). These emblematic trees were confined to the dry, unirrigated areas and continued to either replace wheat farming or be combined with it. Once again, it was the larger farming units in the hands of the remnant of the aristocracy or the liberal, urban middle classes who continued to benefit most from these changes. The small farmers with their limited strips and plots could only really follow in the footsteps of the big producers. Rural population was also at a low level following the emigrations at the turn of the century, which had a limiting effect on productivity in an industry that was not yet mechanised. The day labourer was still the dominant form of employment for the vast majority of Mallorca's agricultural workers. Labour, rather than capital, limited what could be achieved and the countryside remained little altered in appearance until the post-wars recovery of the 1950s. The Civil War (1936–1939) isolated Mallorca from world markets, which meant that the constrained export of processed foods and agricultural produce held back the arboriculture that was beginning to expand. Emphasis went back to bread grains in a desperate attempt to feed the population during those years.

Some attempts were made to consolidate rural population into larger, more efficient settlements rather like the agricultural colony movement of the previous century. The historical geography of settlement in Mallorca is noteworthy for the quite small number of true villages (*llogarrets*) – towns and dispersed settlement being much more frequent. In the south-east of the island, however, it is possible to identify a number of somewhat larger villages (*poblats*) that were consolidated principally as late as in the twentieth century. Their numbers include Calonge, Llombards and s'Alqueria Blanca in Santanyí, Sa Horta and Cas Concos in Felanitx and Son Macià in Manacor. They are inland, away from the coast, and usually on ancient roadways between historic centres. They may look old, but are in fact quite modern creations. Although most had their origins in medieval *possessions*, subsequent subdivisions of estates from the nineteenth century onwards encouraged agricultural expansion and increases in population. The role of landed proprietors was central in allowing the parcelling of the land and creation of villages. In the case of Son Macià (Manacor) the estates were inherited by the twenty-three year

Son Macià *possession*

The new village of Son Macià

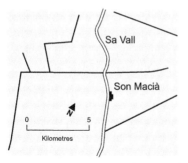

Figure 10.1 Son Macià (Manacor) – a village founded in the 1920s around a possession by the landowning family of Mayol i Gelabert – part of the late colonisation of the countryside. Source: Rosselló Verger (1964) and Garau Febrer (1995).

old Marti Mayol i Gelabert in 1918. In addition to the two local estates of Son Macià and Es Rafal he also owned lands across the island, as well as town houses in Palma (Garau Febrer, 1995: 23). The local *possessions* were soon put up for sale and divided into numerous smaller parcels, including sixty building plots for a village that included a substantial plot for a church (Rosselló Verger, 1964: 381). The initial fifty-seven houses (*solars*) were built on these church lands given by Mayol to the Archbishop of Mallorca during the 1920s. The field boundaries largely determined the morphology of the village as it grew over the next fifty years. It remained relatively isolated until a metalled road replaced the cart track that linked it to Manacor and the coast road in the 1960s. During the 1990s the village flourished, as mainly German second-home owners and permanent settlers bought up the parcels in the valley and their farm houses (Fig. 10. 1).

Industry and transport

In the first forty years of the new century the population of Palma grew by almost 40% largely under the influence of the modernising economy and the continuing decline of agriculture as an employer. The city was in many senses the engine of economic growth. For much of the early part of the last century Mallorca continued to be known as *la terra d'emigració* but by 1930 employment in the secondary sector (manufacturing) had reached a remarkable 50% of the workforce, having risen from about 30% in 1887. In the 1920s of all the provinces of Spain only Valencia had a more rapid rate of growth in GDP than the Balearic Islands. In the following decade (1930–40) Palma grew very rapidly, thanks almost entirely to internal migration. The city, its growing suburbs and their economic opportunities offered new forms of work in the secondary and tertiary sectors caused by the rising demand for goods and services. These began to offset some, at least, of the pressure to migrate overseas to places like Latin America, Algeria and France, which had been deemed so necessary at the turn of the century; the rate was now much slower. New opportunities were being created and many of them were associated with the growth of Palma.

The growing industries may have had their origins in earlier centuries, but they took on a modern aspect in the last third of the nineteenth century, particularly in footwear, textiles and food processing. The loss of the colonies after 1898 had meant that reorientation was necessary for them to survive in a new market-driven environment. Industries like footwear and textiles became increasingly mechanised and productivity was thereby raised. This saw a reduction in the proportion of home working, an expansion of factories and the building of new ones. In both cases, it resulted in the suburbanisation of manufacturing around Palma and the development of new locations in the less constrained environments in the island's smaller towns.

Industrial development in the'20s and'30s was increasingly based on the energy sources of a second industrial revolution – electricity and oil. This was especially true under the dictatorship of Primo de Rivera, whose corporate views encouraged the establishment of monopolies. In Mallorca Joan March founded CAMPSA in 1927 and built the island's first petrochemical works at Porto Pi. Another familiar brand name – GESA – emerged from the joining of the gas and electricity industries in 1927, taking over many of the small private and municipal generating stations (Casanovas, 2005: 48) (Fig 10. 2). While town gas production was largely limited to Palma and was essentially a domestic fuel,

Figure 10.2 An early 20th century (1920) energy generation complex at Zona de Ia Porta des Camp and Can Perantoni, producing gas and electricity. It eventually became part of GESA. Source: Muntaner.

electricity production on a larger scale was to result in the wider spread of industry. Generation was based on coal originally imported from the UK and the mainland – about 30,000 tons by 1921 – but soon supplemented by locally mined lignite from near Alaró, where some 50,000 tons were produced in the 1920s. Large generating stations were built at Portixol next to Palma in 1918 and at Ca'n Perantoni, which were eventually linked to Sa Pobla and the towns of the *raiguer* by the beginnings of an island-wide 10 kilovolt grid (Roca and Umbert, 1990). This was effective in securing the introduction of new machinery and higher productivity in the woodworking and footwear industries in locations such as Manacor and Llucmajor. Although the small firm, often still family based, remained the norm, the number and variety of factories and workshops increased rapidly through the 1930s. In Sóller the number of powered looms rose from 479 in 1920 to 534 by 1940 (Roca Avellà, 2006: 15). The woodworking industry became increasingly concentrated in Manacor, Mallorca's second largest town, a spatial process begun in the late nineteenth century. Sansó has mapped a total of 83 workshops in the town in 1925, consisting of 27 chairmakers, 4 cartmakers, 38 cabinetmakers, 8 general woodworkers and 6 woodturners with a workforce of about 320. By 1935 about half of all the workshops were mechanised thanks to electricity. This gave a distinct urban industrial landscape of woodworking and furniture-making, made up of a multitude of small workshops. The distribution within the town was highly clustered within its old core, producing a pattern of workshops closely networked together and interspersed amongst the housing stock so that work and home were intimately related, as were specialist producers, all dependent upon each other, a pattern that was to continue into the twenty-first century (Escartin *et al.*, 1996; Sansó Barceló, 2009: 117) (Fig. 10. 3).

As relative prosperity rose in the same period, so personal consumption increased, benefiting the clothing, electrical goods and printing and publishing industries, even if much of this increase was class based. In 1919 273 cars were registered in the Balearic Islands, mostly in Palma; in 1925 there were over 2000 and by 1930 over 5000. These, and the same level of increase in goods vehicles, buses and charabancs, were to have a profound effect upon the roadscape of Mallorca. The poorly surfaced roads and tracks would no longer suffice, but modern road building was slow to come to the island, most improvements coming after the 1960s. Visitors in the 1930s still complained of slow and dusty journeys. While many agricultural goods were moved to local markets in the large-wheeled farmers' carts, some of the products used in the expanding food processing industries could now be moved by lorry (Fig. 10. 4). However, because of the state of the roads, the railways remained the most important link between countryside and industrial towns, although because they came relatively late in relation to most industrial development (unlike in the Britain of the 1840s to '60s) the railway was not such a significant factor in industrial location. By 1920 the lines of Ferrócarriles de Mallorca branched out from Palma to link Manacor,

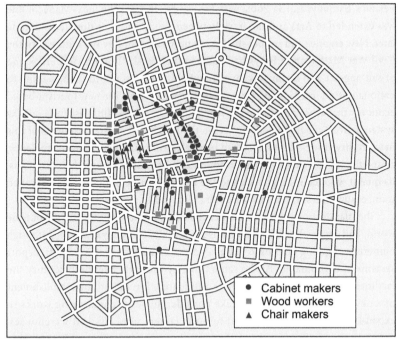

Figure 10.3 The distribution of the woodworking industry in Manacor in 1925. Note the location of the workshops throughout the town in close proximity to residential spaces, a pattern that largely persists today. Source: adapted from Sansó.

Figure 10.4 The road between Bunyola and Sóller in 1930. Although constructed in the 19th century, even by this date it had no asphalted surface, typical of many of the main roads at this time. Source: Muntaner.

Felanitx, Santanyí and Sa Pobla with the capital. In 1921 the Inca–Manacor line was extended to Artà and the track between Palma and Inca doubled a decade later. New engines and rolling stock were purchased, but by the outbreak of the Civil War Mallorcan railways had reached their peak. The post-war era is one of sad neglect and decline as investment shifted from rail to road, leading to nationalisation in 1951, and steam giving way to diesel. The new tourism-based economy that arose rapidly on the scene in the 1950s had little use for a railway system whose geography did not serve the beach and coastal resources of this new activity. The Sóller railway had been electrified in 1929. Its new efficiency, coupled with the attractiveness to tourists of its route through the mountains, its quaint wooden rolling stock and the link to Port de Sóller by old-fashioned tram, ensured its survival into a new era (Fig. 9. 11).

The island's growing industries needed better maritime links with the outside world, and especially with the Peninsula (Pujalte i Vilanova, 2002). Although numerous shipping lines used Palma's port, ship sizes were small and transport systems uncoordinated. At the same time, by the early twentieth century the facilities of the port were themselves antiquated. It required the establishment of a new organisation to undertake considerable civil engineering works to expand the moles and quays and to build larger, more efficient warehouses, creating new maritime townscape features along the city's seafront. For shipping the most significant transformation was the creation in 1917 of the *Companiya Transmediterrania*, the result of state pressure to amalgamate a number of existing firms. This succeeded in three things: reducing costs of imports and exports to the benefit of Mallorca's economy; integrating the islands into a stronger Spanish geopolitical unity and laying the foundation for moving large numbers of visitors, the basis of the burgeoning tourist trade. One tourism activity that benefited from these improvements to Palma's harbour was the cruise liner. Between 1930 and 1935 the number of such visitors rose over three times to more than 50,000 (Barceló Pons and Fontera i Pascual, 2000).

One quite new introduction to Mallorca's landscape in this period was what would then have been called the *aerodrome*, a word that sums up many of the modernist, even futurist, transformations of the'20s and'30s. The airplane had been perfected during the First World War, and soon after machines were developed for carrying passengers in peacetime. Initially flights were short, hopping from the Peninsula or southern France to the Balearics; the first landing was in 1916. Landing strips were close to Palma, the most notable being Son Bonet to the north, originally a private flying club field converted to military use during the Civil War, when Italian and German aircraft were based there. It became the official Palma airport in 1947. In 1940 about 7000 passengers used Son Bonet, numbers that were to rise rapidly to nearly 75,000 by 1950 (Salas Colom, 1992: 53). When it ended in 1945, the Second World War had released onto the market hundreds of military aircraft ripe for conversion, especially DC3s. The way was now open for a massive increase in air traffic from the United Kingdom, France and Scandinavia, and by the mid-1950s, West Germany (Buswell, 2011: Chapters 4 and 5). However, this airport was too close to the growing industrial district of Pont d'Inca and Palma's suburbs, and another site capable of expansion was required. This was found at the military airfield of Son San Juan eight kilometres to the east, where abundant state-owned land was made available. Despite its proximity to the growing tourist settlements such as Coll d'en Rabassa (see below), commercial flights were transferred there in 1959. Three years later more than 25,000 flights were bringing over a million passengers to Son Sant Joan and by 2007 over 9 million passed through the airport. If one technological innovation was to convert Mallorca's tourism industry from 'elitist' to 'mass' in the 1950s it was the aircraft. Naturally expansion in Mallorca was paralleled by similar airport growth in Western Europe, especially in the UK and Germany (Buswell, 2011: 62–3).

The victory of the Nationalists under Franco in the Civil War resulted, in the post-war period, in a new approach to industrial development that was highly centralised and initially bent on self-sufficiency. The consequences for Mallorca were threefold according to Roca: a shortage of raw material and energy resources, a marked reduction in consumer demand, and the rise of an informal sector always in conflict with the central authorities as it sought to supply local needs under an austere regime (Roca Avellà, 2010: 12). For example, the main generating station for Palma was forced to rely on Asturian coal supplies, which were not forthcoming in sufficient quantities. This may have stimulated the island's lignite production from around Alaró, but it was to lead to a near collapse of Mallorca's once flourishing manufacturing industries. The footwear and cotton and woollen textile industries which had boomed in the early 1930s were now short of supplies and isolated from their once growing world markets. However, it was thanks to local initiatives in the network of the original smaller workshops that State interference was largely avoided, stimulating the move towards increased mechanisation and the growth of larger factories. An example would be shoe manufacturing in and around Lloseta. The changing industrial landscape of larger scale factories supplanting workshops shifted markedly by the late 1940s. One of the keys to the expansion of industry was the progress made in electrification. In 1957, on the eve of the first 'boom' in tourism, employment in Mallorca was fairly evenly divided between the three sectors: 39% in the primary sector, 31% in the secondary and 30% in the tertiary sector (Roca Avellà, 2010: 30). The resurgence of local entrepreneurialism clearly subverted the initial intentions of so-called fascist central planning and control, providing an example of a part of the island's landscape as a product of 'unintended outcomes'. Ironically, just as Mallorca had established its first modernised landscape based on industrialisation and stabilised its economy, the onset of the mass tourism industry would lead to the tertiarisation of the economy and quite rapid de-industrialisation.

The early tourism industry

The tourism industry was radically to transform the landscape of much of Mallorca in the twentieth century, and particularly after the mid-1950s, the major topic of the next chapter. Historically, as we have seen, the coast was rarely a place for settlement. From the very beginnings through to at least the late 1700s there were few coastal villages. There were many reasons for this; amongst them was the very real fear of attacks by corsairs and the greater prevalence of malarial environments near the coast. There is also a scarcity of evidence for the islanders being a fishing community for long periods of their history, or even having a fish diet until relatively recently, because of the fear of corsairs. Palma dominated port activity and, to all intents and purposes, held a monopoly in sea-borne trade until the end of the first half of the nineteenth century. Mallorcans were a sea-going people, but much of their history in trade is more a case of the island acting as an entrepôt serviced by Mallorcan and other merchants based in Palma. There were, it is true, a large number of natural harbours around the coast, but most of the early settlements were temporary, often consisting of little more than fishermen's shelters. In a previous chapter we showed that no planned towns of the 1300 *ordinacions* were actually located on the coast. Any late nineteenth century 'colonies' developed on or near the coast were to act primarily as foci for agricultural development. All in all, the island tended to turn its back on much of its coast. Even much of the early tourism industry was not really coastally based, but situated in Palma's hotels and its suburbs.

Lack of space forbids a detailed analysis of the origins of the island's tourism industry; in any case it is well covered elsewhere (for example, Barceló Pons and Fontera i Pascual, 2000; Cirer-Costa, 2009; Buswell, 2011). Our task is to examine the environmental and landscape effects in this early phase before the first 'boom' later in the century. However, it might help if some of the factors leading to its birth were identified. As with any migration (and tourists are simply short-term migrants) a combination of 'push' and 'pull' factors was involved. Amongst the attractions was obviously the island's suitability for tourists in terms of environment and infrastructure and its accessibility via improved transport. The crucial environmental characteristics included safe, sandy beaches, the historic city, attractive inland locations such as the Tramuntana – from about the 1880s European tourists were already falling in love with mountains for holidays – and the island's geology, especially the commercial exploitation of its extensive cave systems. The historical and cultural resources ranged from architecture and archaeology to the anthropological 'quaintness' of the local inhabitants, all of which had proved interesting to Archduke Luis Salvador and his followers in the late nineteenth century. The introduction of the first steam-powered ships between Barcelona, Valencia and Marseilles in the 1830s and '40s had enabled quicker and more comfortable passage. Indeed, until the 1950s the early morning entry to Palma's harbour on the ferry from Barcelona was a major attraction for most tourists.

Nearly all early visitors complained of the lack of hotels on the island. Charles Bidwell, writing in the 1870s, despaired of a decent hotel, noting that there were only *fondas, hostals* and *casas de huespedes* (boarding houses), principally located in Palma's old town. Charles Wood, writing in *Argosy* in 1888, noted a similar shortage (Wood, 1888). By the 1920s a number of significant hotels had been added to Palma's townscape. Domenech's Gran Hotel of 1903[1] (Fig. 10. 5) led the way with its rich decoration of tiles from the Roqueta factory, followed by a series of de luxe hotels built either in Palma or its suburbs or along the coast between Palma and Andratx. These included the Ciudad Jardin (1922), the Mediterráneo built on

Figure 10.5 The Gran Hotel. This was Mallorca's first modern hotel, built in 1904 in the Moderne style. In many ways it marks the beginning of the commercial tourism industry beginning a phase of luxury hotel building that lasted through the next two decades.

Figure 10.6 The Hotel Mediterráneo opened in 1923 and expanded in 1927. It catered principally for overseas visitors and offered luxurious accommodation. This first generation of hotels was located primarily in Palma or to the east and west of the city. Source: Muntaner.

the finca Barra d'Or (1923) (Fig. 10. 6), the Hotel Victoria (1928) on the finca Sa Sabater, the Principe Alfonso in Cala Major, the Hotel Playas de Peguera (1928) and Malgrat (1932) (Seguí Aznar, 2001: 41). Perhaps a different but iconic hotel of this period was the Hotel Forementor, built in 1928 in a quite different kind of location. This was developed by the Argentinian Adàn Diehl (1891–1952). He, with others, acquired land in the Formentor peninsula overlooking the Bay of Pollença, but it was almost inaccessible by road, so that building materials had to be brought in by boat. It had an isolated location that appealed to Diehl's romantic notions, stemming from his connections with Argentine poets and artists. However, the hotel was a strictly commercial venture. It was not aimed at the artistic colony that had grown up at nearby Port de Pollença (Goldring, 1946) but at an upper-class elite looking for privacy. Slowly the number of hotels increased to over 130 by the mid-1930s.

Hotels were important as the original resource for tourism, but equally significant in this period before mass tourism was the beginning of new settlements dedicated to visitors and built on greenfield sites. They were aimed at a highly selective type of property ownership and seasonal renting for longish periods, often taking advantage of Mallorca's ameliorated winter climate. There was a strong element of the romantic attached to these settlements, drawing on the artistic sojourners, often of a more bohemian persuasion, who had already started to drift to the island attracted not only by its climate but also by its very low cost of living. To this group might be added the ex-colonials (of Spanish, British and French Empires which were declining by the 1930s) who were able to find cheap property and cheap servants, enabling them to maintain a lifestyle not too dissimilar from that which they had left behind. At the same time as the population of Palma grew, so pressure on the old city built up, and the rising bourgeoisie began to look outwards to suburban living, as was happening in so many north European cities in the 1920s and'30s. So we can detect a series of linked spatial processes going on at the same time, some related to the burgeoning tourism industry and others to wider urban spatial changes – punctiform hotel building in Palma and in the bays and beaches elsewhere on the island, but especially along the south coast, more particularly in Calvià, suburban development on Palma's edge and new tourist settlements on greenfield sites, especially in the east and north of the island. Seguí describes these new forms of urban development as *ocio* or leisure settlements (Seguí Aznar, 2001). The processes involved reveal Mallorca's adoption of some European planning ideas prevalent at the time, producing landscape elements that echoed patterns in Britain, Germany and France. One influential designer was the Italian-born, Argentina-raised, Paris-trained architect–planner Felipe Bellini.

La Ciudad Jardin de Palma was a private venture begun in 1918 to combine residential suburban development and the needs of tourism. Land was acquired at Can Matorell and Son Moix in Coll d'en Rabassa. The plan involved a mixture of plots set within the symmetrical structure of a late eighteenth-century garden, with a central avenue and plaza and tree-lined roads. A series of polygonal blocks was to contain two to eight houses per block. It was an eclectic idea covering sea-bathing facilities, a hotel, summer homes with the air of a 'garden city'. By the twenties beach culture was beginning to turn away from the medical and social prohibitions that were imposed in the nineteenth century, when sun and sand were said to be bad for the body, to sunbathing, even nudity, and hours spent on the beach. The beach began to take on a new socially constructed meaning at about this time. It changed from the utilitarian seashore of an earlier generation to the pleasure beach; it became for many of the visitors to Mallorca in the'20s and'30s 'a stage, a playground, a site for edification, sexual gratification or simply easy living' (Baerenholdt *et al.*, 2004: 51). The beach became the geographical starting point for many new urban developments. The Mediterranean 'season' was no longer the cooler period from October to April but high summer (Lenček and Bosker, 1999: 198–204). In the United States seaside settlement planning was very much the product of the Art Deco movement, but in Mallorca it was clearly influenced by the garden suburb notions of Unwin and Parker prevalent in Britain at a slightly earlier period. It can also be seen as containing the seeds of Ebenezer Howard's planning ideas with its development separate from Palma's nucleus. [2]

Other urbanisations designed on similar principles also date from the 1920s and'30s, many located at a distance from Palma, thus aiding the geographical spread of tourism. These included San Antonio de la Playa at Can Pastilla, Palma Nova, and Portals Nous in the south and Playa de Alcúdia in the north. Their planning took account of the local environmental conditions, with construction on greenfield sites often on former farmland on the often neglected estates. They tended to be limited in size so that they did not seriously affect the environment nor transform the island's spatial relationships. They were intended primarily for summer occupation by the middle classes. Usually they had an axial road or a square orientated towards the beach. Housing, often Ibizan in style, consisted of single-family dwellings, but there would also be a hotel and sports and recreation facilities. The grid pattern of the streets reflected the medieval traditions seen elsewhere in the island from the ordinances of 1300. Three good examples of this movement are Cala d'Or, Santa Ponça and Palma Nova.

Cala D'Or, in Santanyí municipality, was developed by Josep Costa Ferrer (1876–1971) who acquired land in 1933 made up of the fincas of Ses Puntes and Ses Calbresas, and included three *calas* – Llonga, Ses Dones and Gran – which were to prove key to its later expansion (Fig 11. 3) . Initially the urbanisation Costa and his architect Bellini planned had a plaza-like open space and a park. Indicative of the kind of market such developments were aimed at were theatre designers, film directors, Bellini himself, the wife of Rudolf Valentino and many artists. In 1934 the Belgian Van Crainest obtained permission to build a small hotel of 30 bedrooms in Caló de Ses Dones. His artist compatriot, the painter Vanbergh, acquired land adjoining Cala Llonga where he built the first marina. Later the finca Ses Quarterades would form the basis of Ses Marines urbanisation on land between Cala Gran and Cala Esmeralda, although development ceased with the outbreak of the Civil War. Its plan, also by Bellini, differed from the two previous urbanisations by having a much less formal layout, giving irregular building plots with a large park of pine trees in the north next to Cala Esmeralda. The three initial settlements were to coalesce into the present day but unhistorically named Cala d'Or (Cirer Costa, 2010: 276–8).

Another example is Palma Nova, begun in 1934 by promoter Llorenç Roses, a Puerto Rican by birth, with architect Josep Goday i Cassias. Part of the original finca de Ses Planes between the Andratx to Palma road and the sea, it was to be given over to a model garden town, again relating the architecture and layout to the countryside and the sea. The main axis would be from the natural promontory, where a large hotel would be located, to a central plaza, itself placed on the highest point of the finca, a place for *grandes festivales civicos*. Land would be divided into 1000m^2 plots, some for houses with gardens,others for sports facilities, public buildings, a hotel, a church, a school and local shops. There were to be strict guidelines for house building, each scheme to be submitted for approval by the promoter or the residents' committee, nothing taller than '*una planta baja y un piso*' (a ground floor with a flat above) except for hotels and public buildings. While the basic design style was to be Mallorcan, elements of *la masia catalan, cartijo andaluz* or *la casa menorquina* would be allowed. Roads would be six or eight metres wide with pedestrian ways of four metres. The town would be divided into two zones: to the east *casi llana,* and to the west a wooded area. Two main roads twelve to fourteen metres wide would link the area to the main Palma road. There was to be a *paseo maritima* following the coastline. However, as was so often the case with this type of settlement in Mallorca, when Palma Nova came to be built the road layout was adhered to, but the building designs were not.

El Terreno, to the east of Palma around the bay, was different from the 'garden suburbs' described above. It began life as a zone of summer houses for the middle classes and small proprietors of Palma (Fig. 10. 7). It developed in the 1880s on land sold at ridiculously low prices as a series of small houses (*casetes*) painted white, yellow and blue spreading out from Palma's fringe, especially after the demolition of the city's walls. El Terreno began to be transformed, thanks to the influence of foreigners like the English in the winter season, renting houses, opening a Protestant church – *un centro social y eclesiàstic angloamericana* – to which was added luxury hotels like the Hotel Mediterráneo in 1923 and the Victoria Regina in 1928. John Walton believes that the Anglo-American sojourners saw El Terreno as a kind of hedonistic paradise, eventually serving an international community of about 500 (Walton, 2005: 183). From about 1925 the area began to be converted into a residential zone, particularly between 1930 and 1936 with the construction of blocks of apartments, losing its original identity as it became absorbed into a wider Palma. According to Bartomeu Barceló Pons, the construction of the Paseo Maritimo at the end of the 1950s finally changed the nature of the area (Barceló Pons, 1963). The former main road to Andratx, now known as Calle Joan Miró, contained the major shopping outlets and services, with a focus on Plaça Gomila, but its through traffic function was replaced by the Paseo Maritimo. Calle Joan Miró divided the area into two very different zones – one towards the sea, the location for high quality family homes often architect designed, one of which, Cas d'es Palts, has been attributed to Gaudí. Barceló identifies a number of housing styles, beginning with detached houses in gardens, set in parallel streets, built between about 1880 and the turn of the century. In the second phase from the 1920s came *la casa de planta y piso,* often decorated, now divided into five or more apartments. Once it had taken on a proper suburban function and landscape, El Terreno is better dealt with as an adjunct of the city, as Gabriel Alomar realised in his plans for Palma of the 1940s (see below).

The Civil War (1936–1939) seriously upset the ongoing development of the tourism industry. British and American residents – including Robert Graves – were advised to leave Mallorca and tourist companies and cruise liners took the island off their itineraries, despite the neutrality of most European countries. During the war Mallorca was primarily in support of Franco's fascists and the only invasion by Republican forces, via Porto Cristo, was easily repulsed. Mussolini's not-so-covert air force used Mallorca as an offshore aircraft carrier from which to bomb the Republican strongholds in Barcelona and Catalunya. Naturally the island economy suffered severely both during and after the war. Little new building took place, large areas were occupied by the military, and above all, food supplies were desperately short. Isolationism and a closed economy became the watchwords of the new fascist government of Spain after 1940, leading to a period of

Figure 10.7 El Terreno, to the west of Palma, developed initially in the late 19th century as a location for the summer houses of *palmasanos*. By the 1920s it had become colonised by overseas visitors and had a somewhat bohemian reputation, noted for its bars and its nightlife. Source: Muntaner.

deep recession, price fixing and rationing, all of which encouraged smuggling and black marketeering, well suited to all island of many isolated bays and beaches. However, the Franco government was insistent that the industrial developments begun in the 1930s should be expanded; tourism was not to be encouraged as a serious economic activity. It was not until the 1950s that the economy was opened up, particularly after the United States provided $62 million of credit designed to lead to more inward investment to support Spain's non-convertible peseta. Large-scale inflation soon followed, accompanied by devaluation, thus opening up the floodgates to overseas visitors – and more importantly, to tour operators and overseas holiday companies – who found their harder currencies bought large amounts of holiday pleasure – and investment – especially after the approval of the Stabilisation Plan of 1959 (Alzina *et al.*, 1994: 390). The landscape effects of this first 'boom' will be examined in detail in the next chapter. Meanwhile, tourism in the period before the 1950s was not insubstantial, especially for visitors from the Peninsula. Cirer gives considerable prominence to this pre-war(s) period, ascribing to it 'boom' characteristics, largely thanks to the growing affluence in mainland Spain, particularly in Catalunya. Much improved ferry services from Barcelona and Valencia aided the increase of tourist numbers from 36,000 staying in 19,000 beds in 88 hotels in 1930 to more than 90,000 tourists in 1935 (Cirer Costa, 2010: Chapter 10; Buswell, 2011: 46–7). Cruise liners brought 50,000 day visitors by 1935. Naturally recovery was very slow thereafter, but 55,000 tourists came in 1945, 610,000 in 1950 and 1. 5 million in 1955 (Ginard, 1999: 13).

Tourism in this period is in some ways an extension of the processes begun in the late nineteenth century, but on a considerably expanded scale, and this is reflected in Mallorca's landscape. Four key factors help explain its growth: improvement in international communications, the availability of cheap land on the coast largely derived from the break-up and sale of ancient estates, the building of specific resorts for leisure and tourism and the continuing attraction for tourists of the dominant urban centre, Palma.

The growth and planning of Palma in the first half of the twentieth century
The creation of the *ensanche* by the demolition of the city's walls at the turn of the century led only to slow urban development of the new space but, as we have seen, it did not solve many of the problems of overcrowding, sanitation and internal distribution and movement that had become acute by the early 1900s. This increased demand for a *planning* approach that addressed some of these issues. In order to appreciate the way in which the townscape evolved in the first half of the twentieth century it is necessary to outline briefly some of the plans put forward.

Planning in Palma was not, of course, new; nearly all of the layout of previous city structures had been planned to some extent, but the general appearance of the city's form was the result of rather random growth in the various districts; there had been little attempt to plan the city overall. Mallorcan architects and engineers were clearly *au fait* with current ideas in planning by the turn of the century. As in the case of the seaside settlements described above, the work of British and American planners was especially well known, another example of the island's intellectual connectivity with a wider world, far from being isolationist.

A more comprehensive approach based on urban theory was very much the product of the late 1800s, but it became clear that to relieve inner area pressure and create a more attractive living environment, new lower density suburbs would have to be built – what became known as *eixamples*. Some of these had developed in the nineteenth century, such as Santa Catalina, founded in 1869 and Soledat (built on three *hortes*) ten years later, suburbs that in 1900 had, respectively, populations of 7400 and 1240. These were working class areas related to industrial development, but many other informal urbanisations had also arisen. Around the bay to the west El Terreno, as we have shown, had developed as a zone of very low density middle class housing evolved from its seaside, summer residence origins, as described previously.

The first comprehensive plan was that of Bernat Calvet, drawn up in 1901. He divided the city into two parts separated by the line of the former Riera watercourse, and proposed two suburban expansions in the area beyond the *ensanche*, but in reality his ideas focused more on the need to improve radial routes coming into and out of the city rather than tackling the more pressing problems of the congested, inefficient inner, historic core of Palma. To improve other aspects of the infrastructure he also proposed four new markets, libraries and other municipal buildings and a site for the Ramon Llull Institute, but as Carme Ruiz Viñals points out, his *eixample nomes es dibuixa però no s' estructura* ('designed but not built'), principally because the city authorities did not have the political will to enact it (Ruiz Viñals, 2000: 104).

In the next two decades Jaume Alenya in 1912 and Gaspar Bennàzar in 1917 drew up alternative plans. Alenya's plan was largely concerned to improve links between the upper and lower parts of the city, but as with Calvet, little progress was made. Bennàzar's plan, on the other hand, was approved by the Ajuntament. He was asked to tackle the problem of improving the old urban core, but without destroying its architectural heritage, a problem the modern movement was to face later in the century. He proposed a series of wide avenues connecting all parts of the central city including linking Plaça de Cort with Born and Sta. Catalina and another joining Sta. Eulalia with El Temple. In the end this was another plan that was never fully realised.

The question of traffic congestion was addressed in 1925 by Guillem Forteza Penya (Seguí Aznar, 1999). He noted that the main pinch points were where the roads from the north and north-west met at Raconada de Santa Margalida and clashed with traffic generated by the railway station, where roads from the east and north-east met at St. Antoni Gate, and thirdly, where port traffic had to fight its way through the city centre. Like his predecessors, Forteza focused on the construction on new throughways – a circular road from Conquistador to Oms that would improve links between the high town and the low town, and two Gran Vias, one connecting the city centre to Raconada de Sta. Margalida and another 15 m wide joining the second congested area, Porta St. Antoni, to the centre. His plan also called for the then quite radical idea of zoning, derived from early American planning theory, with separated areas for residence, business, industry and leisure pursuits. Surprisingly, he also advocated segregated housing on a class basis.

The demolition of the city walls facing the sea presented an opportunity to address east/west communications by building a *passeig maritim* going through to Porto Pi and the expanding suburb of El Terreno, and providing Sta. Catalina with an esplanade. Forteza, and earlier Bennàzar, both developed this idea but it was not confirmed until the plans of the 1940s were approved, and even then not opened until 1959.

One technological advance that assisted the growth of the city after 1900 was the introduction of the first mass transit system, the tram. Mule drawn trams, said to have been acquired from Liverpool, had been introduced from 1891, but were quickly replaced by electric trams from 1914. By 1916 all animal power had been superseded (Fig. 10. 8). The tram network not only eased movement within the inner area but, more importantly, linked the expanding suburbs to the city centre. Table 10. 1 below shows the date at which more distant places like Genova,

Figure 10.8 By the 1920s Palma had quite a dense network of tramways. Originally drawn by mules, they were replaced by an electric system in the century's first decade. As in other European cities the trams were important in opening up the suburbs to commuters. By 1930 the population of Palma had reached about 88,000. Source: Muntaner.

Establiments and La Soledad were connected. At its greatest extent more than 5000 km of lines were laid down, but by the late 1950s the trams had disappeared. As in other western cities this was largely due to changes in urban morphology, the increase in the number of new settlements, a lack of investment and, of course, competition from the internal combustion engine and car ownership (Bibiloni Rotger, 2002: 10).

Table 10. 1 Tram routes in Palma.

End of route	Began	Ended
Porto Pi	1916	1955
Circular	1916	1954
Can Capes	1916	1959
Es Molinar	1920	1941–8
San Roca	1921	1958
Genova	1922	1958
La Soledad	1922	1958
Establiments	1926	1959
Coliseu	1929	1958
Port	1934	1935

Source: Bibiloni Rotger, 2002: 59.

Lack of space prevents a detailed analysis of the contribution of architecture in the first forty years of the century to Palma's townscape, but the College of Architects' guide book gives some excellent examples (Lucena *et al.,* 1997). Two key movements were the *moderne,* equivalent to art nouveau, and the modernist movement, with its origins in Germany.

Gabriel Alomar Esteve (1910–97), one of Spain's leading architect/planners of the twentieth century, had studied in the USA at MIT and brought pioneering ideas to bear in his plans for Palma in the 1940s. They were eventually to produce much of the modern townscape of Palma that we see today, often by default (Alomar Esteve, 1950). The plans were based on an attempt to remedy at last some of the issues in the city's layout, and its social geography, which earlier planners had described but had been able to do little about. Three objectives were identified: to improve the development of the *ensanche,* to bring the informal suburban settlements that had grown up since before the turn of the century into the overall plan of the city and to modernise the ancient core of Palma. In many ways this last reform was the most pressing, partly for reasons of public health, which in turn reflected an orthodoxy that had reigned over much of northern Europe a century earlier, namely that many of a city's ills could be eased if inner area densities could be reduced by clearing out population from the medieval lanes.

By the 1940s Palma's growing tertiary economy also required the changes of land use necessary to provide the space for modern urban functions. This would involve the straightening and widening of the street network to improve accessibility and internal movement, together with the provision of new spaces for offices, shops and other business premises – in all, the creation of a proper functional central business district. Alomar's method was to divide the city into twelve districts, some of which, like parts of Calatrava and Puig de Sant Pere, would be largely demolished to make way for new roads and housing. They were poor, overcrowded districts and little opposition could be expected. However, such areas were of great historical importance, containing many significant architectural treasures. Luckily the political costs of redevelopment defeated the city fathers. Such areas were later to become the cornerstones for the conservation of the historic city, with new uses being found for old buildings from the late twentieth century onwards, producing some of the most expensive and sought-after property as inner city living took on a new cachet. In the event, only two inner areas were subject to radical reform through Alomar's 1943 Plan (Alomar Esteve, 1943). These were the creation of a new area for the city's central retail market in the new Plaça d'Olivar, and the Calle Jaume III. The second of these proposed a new, wide, but very urban street running through an old historic quarter from the north-west of the Born. It was to be lined with high-grade arcaded shops with expensive apartments, offices and hotels built above. Jaume III formed the axis of a triangle of development stretching outwards, towards what was later to become the Paseo de Mallorca.

The architecture of much of the area was to reflect values current under Franco: permanence, massiveness and authority. Alomar's third reform, and perhaps his most dramatic, was his proposal for a high-speed highway located between the city and the sea, capable of redistributing traffic to east and west and complementing Calvet's boulevard along the lines of the old city walls. Although not built until the late 1950s and designed by Roca, this new landscape feature also had the psychological effect of separating Palma from its raison d'etre, the Mediterranean, appropriate at a time when air travel was taking over from the ferries, and when the new tourist economy was taking off.

Change and growth

Modernisation, then, was the key feature of landscape change in the first half of the twentieth century, reflected in the patterns of the development of industry, business services and retailing in the larger urban areas. In agriculture the considerable expansion of the area under cultivation saw crop yields increase, leading to a rise in the export of the island's farm products. This gave the landscape of the rural parts of the island an invigorated aspect. With electrification, even if it was highly localised, the manufacturing industry expanded its exports, especially in furniture, footwear and processed foods. By the 1950s Mallorca could lay claim to at least a partial industrial landscape. Planning from the centre became a key part of this process as new roads and buildings were laid out, especially in the capital city, Palma, where theoretical ideas of urban development, imported from Great Britain and the United States, were adapted to Mallorca's social and physical

ANDRAITX. S'Arracó. Playa de San Telmo e islotes "Pentaleu" y "Dragonera"

Figure 10.9 Sant Elm, photographed in 1915, epitomises the 'paradise on earth' that many early visitors saw in Mallorca – a deserted beach surrounded by pine trees, an unspoilt landscape typical of coastal Mallorca, which was to be transformed utterly by the impact of the mass tourism industry: from Paradise Found to Paradise Lost in less than 100 years. Source: Muntaner.

environments. The creation of a city that amalgamated its historic core with well planned suburbs would form the engine for much of the transformation that was to come in the century's second half. At its heart would be a tourism industry which began its slow progress to 'take-off' in the 1920s with many new hotels and garden-village settlements for the adventurous middle classes, who saw in Mallorca's 'undiscovered' pristine beaches a kind of paradise (Fig. 10. 9). It would go on to create a built landscape of coastal resorts that in many ways were a model for such developments elsewhere in the Mediterranean, but on a scale that would have been unthinkable to the builders of the 'garden city' resorts.

Mass tourism and the landscape – Mallorca 1955–2011

Una de les transformacions mes profundes, fruit de l'impacta del des d'envolupament del turisme de masses ha estat el radical canvi fisic de l'espai illenc, que resulta de la mutació d'extenses superfícies de sol agrari en sol urba ... un impacte mes important sobre el paisatge natural.

(One of the most profound transformations, the product of the development of mass tourism, has been the radical change in the structure of island space, which has resulted in the extensive and sudden change of agricultural into urban land, a most important impact on the natural landscape) [Pere Salvà i Tomàs, 1985]

To the urban cultures of the industrialised west the coastal strip where the land meets the sea is not just any kind of place. The beach plays a significant role as an attractor of fantasies and desires, one of those leisure spaces specially designated for the purpose of pleasure and physical gratification for modern city dwellers. [Baerenholdt, Haldrup, Larsen and Urry, 2004]

Sens dubte la història de la segona meitat del segle XX a Mallorca tendra un nom amb tots els atributs del deus. Un nom omnipresent, omnipotent i totpoderós, aquest deu es diu Turisme.

(Without doubt the history of the second half of the twentieth century in Mallorca has a name with god-like attributes. An ever-present name, omnipotent and all-powerful, this god is the god Tourism) [Jaume Santandreu, 1990][1]

A brief overview

The last sixty years have seen such a transformation of so many aspects of life in Mallorca – and landscape foremost amongst them – that it is difficult to comprehend; it is surely a case of the 'shock of the new'. The motor that has driven these changes – and they continue – at such high speed and so pervasively – has been holidays for the masses of Europe. In the previous two chapters we saw these processes emerging; the shifts were significant but relatively small and limited in their effects. The 1930s had witnessed the beginnings, the 1950s a second renaissance, but both of these phases tended to add to the economic and social life of the island rather than swamp it, nor obliterate so much of what had gone before, nor change rapidly what people did for a living or the appearance of the island. In the fifteen years or so before 1973 the rate of change in Mallorca's coastal landscape was so dramatic it was difficult for local observers to keep up with the changes; the French geographer Colas, quoting government statistics, estimated that in 1964 a new hotel was being opened in Mallorca every two days, four hours and forty-eight minutes! (Colas, 1967). Long-deserted coves suddenly acquired a new status, landowners – large and small – sold or leased the land their families had held for generations and their sons and daughters no longer followed their parents into agriculture or stayed in the village or the country town. They flocked from the Pla to the new boom settlements at the seaside looking for jobs as waiters, bedmakers, cleaners, barmen, receptionists and cooks. The coast long avoided by Mallorcans for fear of corsairs now became a valued commodity. A society that had only one sizable town now saw arising on its shores veritable Manhattans. The population rose from 363,000 in 1960 to 561,000 in 1981, but that was just the people counted as residents; the 'real' population in the summer was clearly much higher; where there had been hundreds and thousands there were now millions. By 1995 over 6 million visitors came to Mallorca; by 2011 this had increased to nearly 10 million. In August 2011 alone more than 3. 4 million descended on the island (Diario de Mallorca, September 11, 2011).

In the longer run the changes wrought by this new tertiary activity – what a century earlier Amengual had called *l'indústria de foresteros* (Amengual, 1903) – the industry of foreigners – was not confined to tourism alone. Initially investment had been in the construction of new hotels, apartment blocks and the infrastructure of resorts generally, but soon the economic system widened to embrace all the services that supported the visitors, everything from banking to laundries. The multiplier effect soon began to then move into another cycle to provide services for those that serviced the tourists. An island with only poor agriculture and a population that was small relative to the number of visitors soon proved unable to supply what was required, and this in turn stimulated the importation of a whole range of *materiel* and people. These imports manifested themselves in many ways, including capital and revenue from overseas tour operators, particularly from Britain and later Germany, who sought to invest in hotels. Workers of all kinds came from the Peninsula, especially construction workers from Catalunya, Murcia and Andalucía. Foodstuffs came from wherever they could be had, fuel from Europe and the Middle East to increase the electricity output of the island's generating stations, and at times of drought even water came from the Peninsula in tankers. Altogether this is an excellent example of the economist's *circular and cumulative causation.* [2]

At the same time the economy and the social organisation of the host, Mallorca, had to reorientate itself, the social effects of which have been complex and longlasting and which, regrettably, cannot be considered in detail here. Nonetheless they do underlie many of the landscape shifts we are about to examine. Demography, education, language and religion are but some of the social elements that were subject to massive change. Often at the heart of things was the tension between Mallorcan and incomer, the new settler as well as the tourist. The increased demand for labour was not simply a matter of attracting more workers to the island, but of accommodating and coming to terms with their apparent cultural differences. And halfway through our sixty-five year period the whole political basis of Spain changed from a virtual dictatorship to a new form of social democracy, and for the Balearic Islands as a whole, a large degree of autonomy and a *Parlament* based in Palma.

In geographical terms two processes are worth identifying in this preliminary overview: the urbanisation of the coastline together with the continuing and increasing primacy of Palma and the counter-effect on the agricultural heartland of Mallorca, which rapidly lost its historic *raison d'etre*, and many of its people. Environmentally the impact of mass tourism has been, and continues to be, complex and its long-term effects not fully understood. Clearly the population, both permanent and seasonal, have increased while the size and physical structure of the island have remained constant, leading to inevitable and at times intolerable stress on natural systems. The land has been mined and quarried to provide building materials, the water supplies both above and below ground have been brought to the point of collapse in some dry years. The waste material generated by so many visitors may soon overwhelm any system devised for its management; road building and traffic generation together produce much pollution, especially on the crowded south coast and in Palma, where Mallorca has more cars per head of population per square kilometre than anywhere else in Spain, inevitably leading to worsening air quality. Energy demand has risen so dramatically that the island's own system can barely cope, so that it had to be connected to the mainland by a 400 kilowatt capacity undersea electric cable in 2011; a twenty-inch natural-gas pipeline was opened in 2009. The gas will be used to generate electricity in Son Reus's recently upgraded power station. In addition to this new plant the island is in effect being 're-wired' and older, low voltage transmission lines are being replaced by larger pylons with a more noticeable presence in the landscape. The benefits will be felt in the control of the island's carbon footprint.

Although this chapter is conveniently titled 'landscape and mass tourism', it is already evident from this introduction that tourism alone is not the only variable producing dramatic landscape change. It will also be necessary to consider the role of other service industries, manufacturing and distribution, urban development and housing and the general movements in a very affluent society that is already on the way to being post-touristic. Tourism may be the engine, but it now pulls in its wake many other carriages.

In spatial terms a final point to make is that the effects of mass tourism are not as geographically pervasive as might be expected. Mallorca is quite a large island, and so far tourism has been primarily located on the coast, with a limited number of visitors' honeypots inland. We will show that at the beginning of the twenty-first century there has been a significant growth of second homes in the countryside, together with an expansion of some of the larger inland towns especially in the south. There is also some evidence to indicate that in many areas the footprint of the visitor or the new resident has been slight. Much of the centre of the island, its western highlands and its eastern hills retain many of the visual landscape features of the old, inalienable, Mediterranean Mallorca that has taken centuries to evolve slowly. In a limited number of areas, when most of the summer's visitors have gone home, they seem to breathe a sigh of relief and revert to their former selves.

The landscape of mass tourism on the coasts of Mallorca: some theory and some examples

It has been conventional to see the development of tourism in a series of stages, each one contributing distinctive elements to the landscape, some of which will have survived, others being swept away as part of the process of renewal and change. In other words, change in the landscape of tourism is very like the change in the landscape as a whole. It is, in the case of Mallorca, possible to see an historical pattern of infrequent visits utilising local accommodation facilities at the beginning, poor as they may have been. This was followed by the building of a punctiform distribution of hotels for a tourist elite with a focus on the principal urban place where a range of services existed. Later, resorts would emerge, perhaps around these hotels or on greenfield sites that exploited new seaside environmental resources. In resource-rich areas or in popular locations some of these resorts would have expanded and possibly coalesced. Large specialist resorts may then have been constructed. By this time much of the expansion of the built tourism infrastructure would have become merged with a general process of island urbanisation, with many settlements performing dual functions. Finally there might be decentralisation from the original major cores by both suburbanisation and counter-urbanisation, followed by sporadic, small-scale growth in inland areas, the conversion of farms and fincas, agrotourism and the second homes of visitors and locals – the urbanisation of the countryside. [3]

Clearly, a common denominator in terms of landscape has been the fact that tourism has been synonymous with urban development. It would be useful, then, to begin with a little light theory on the process and patterns of tourist settlements and their townscapes, finally tracing the actual historic developments in Mallorca itself.

There are two starting points, which we described in previous chapters and need not repeat in detail here. First was the centrality of Palma in the process. This was the point of entry, the place of accommodation and the centre for exploration. The second would be our search for pre-urban nuclei at the site of seaside environments. These prove to be scarce in Mallorca for historic reasons. At most there might be temporary fishermen's huts, isolated nineteenth-century agricultural colonies or perhaps small harbours around which tourism might begin. Beach culture in Mallorca seems to start in earnest in the 1920s, when it was linked to notions of fitness and health. By the 1930s sunbathing and swimming became

Figure 11.1 This hotel built in the 1920s in Port de Pollença occupies a calm and peaceful place, which after the mid-1950s was to be lost to the massive expansion of the tourism industry. Source: Muntaner.

more pronounced, following both German and American leads. This was also an era of growing bohemianism, reflected in a need to escape from bourgeois ideas of correct behaviour, possible in isolated bays and coves away from urban strictures, distant from Palma. The wealthy also began to demand this kind of isolation away from the prying eyes of the press, but with the high level of service they could afford. All of these types of tourism activity were reflected in the morphology of the resorts. Palma had its hotels de luxe located close to its central business district or on the bays and headlands in close proximity to the city. On the other hand, places like the bays of Pollença and Alcùdia provided the beach environments for both bohemian aesthete and industrial or political mogul (Fig. 11. 1). Douglas Goldring, writing in the 1920s, describes the pleasures of the northern beaches[4] while the Argentinian entrepreneur Adan Diehl developed Hotel Formentor, originally for his artistic friends but later for much better-off clients from the world of business (see Chapter 10).

Andriotis has produced a useful taxonomy of the morphology of resorts based on Butler's well-known six stage tourist area life cycle model of 1980 (TALC) applied in a matrix to ten elements of morphological change: beach width, residential areas, farming land, road network, tourism, businesses, lodgings and infrastructural facilities, second homes, ribbon development, architecture and overall morphological transformation (Andriotis, 2003 and 2006). It might be argued that it was the physical geography of a coastal landscape characterised by the beaches so essential to the seaside culture of the 1930s that led to the kinds of resorts we are most familiar with today. This was certainly true for the growth of hotel-based tourism that developed rapidly in the early 1950s, particularly in the area between Palma and Andratx in the municipality of Calvià. The second half of the twentieth century, however, was to witness a much more aggressive form of coastal urbanisation related to tourism (Salvà Tomàs, 1990). The physical basis for this was, first, access to a substantial beach, narrow but long – rather different from the cove and *calas* beaches which formed the focus for the garden city developments of the'20s and'30s. Often a *torrent* would be another physical element because where they decanted into the sea, a beach would form. These intermittent water courses had to be controlled artificially in concrete canyons in order to protect the incipient settlement from the periodic flooding so common in Mallorca. Secondly, the site had to be capable of expansion to include not only other hotels but apartments, shops, discos, cafés, bars and restaurants, all of which 'fed' on the hotels. Thirdly, it needed good access to the airport and other parts of the island – the first for ease of transfer by coach or taxi, the second to facilitate side tours and sightseeing to attractions such as caves, markets, mountains and historic sites – also by coach. Translated by economics in which economic rent or income was dependent upon access by maximum number of tourists, this would result in such a point – the Peak Land Value Intersection (PVLI) – occurring where access to roads met the front-line in close proximity to the beach (Fig. 11. 2). The hotels themselves would rarely be on the beach itself, but would be protected by gardens,

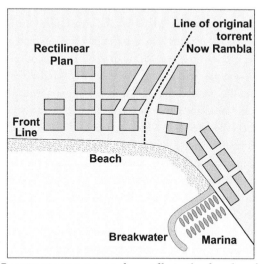

Figure 11.2 Resorts grew up extremely rapidly in the decades after the 1960s. Although planned in theory such settlements, in fact, were subject to ruthless development by local and inward investors over whom local authorities had little control. Most followed a grid-plan layout with the peak land value intersection where the main access road met the coast producing a 'Manhattan' landscape of high-rise hotels and apartments – a wall of cement – along the high value seashore – the so-called 'first line'. In the 20 years after 1980 many resorts developed large marinas, a new addition to the townscape.

swimming pools, tennis courts and other open spaces; later a formal promenade would be built directly by the beach, usually by the municipality rather than the hotel developers. Behind the hotels on the front-line would develop other land uses such as bars, cafés, beach-related shops, etc. The street plan of these urbanisations in Mallorca in the '60s and '70s was nearly always a gridiron. There are probably two reasons for this: an economic one based on maximising income from space, and a cultural one in that this kind of lay-out has a long history on the island, as we have seen. The grid of streets would be laid out in advance accompanied by the usual services of water, electricity, street lighting and a sewage system. Infill would gradually take place as the resort expanded.

This kind of grid plan urban development, although common, was not the only model in the thirty years from 1955. Building around small beaches protected by rocky headlands continued. The valleys between the headlands would initially be avoided (see above) with a much more curvilinear street plan being set out on the higher ground (Fig. 11. 3). Other resorts developed a more conventional town plan based on a pre-existing settlement established for purposes other than tourism. These would include holiday centres evolving from harbours, small ports and old agricultural colonies. Some urbanisations emerged looking something like the housing estates familiar to the British middle classes, but built for seasonal and permanent occupation and sometimes given rather bizarre road and street plans

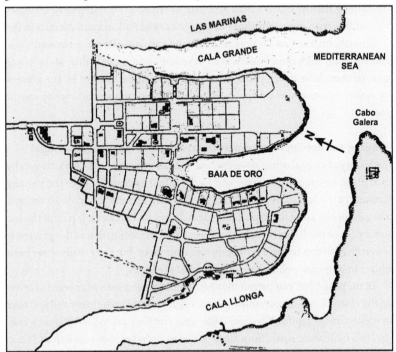

Figure 11.3 Plan of Cala d'Or, a resort begun in the 1930s along garden city lines but massively expanded in the 1980s and '90s. Source: Buswell.

such a circles and squares, upon which would be built speculative detached housing with spacious gardens, thus following a tradition started in the early 1900s.

On this basis it is possible to propose a six-fold classification of tourist settlements from the '60s, '70s and early '80s that is at the heart of many of Mallorca's townscapes:

- The self-contained hotel resort, such as Formentor;
- the resort suburban to Palma performing peri-urban functions for the city as well as the visitor. Examples here would include Platja de Palma, an expanded El Torreno, and Palma Nova. Indeed, many consider the urbanised 12 km beach front to the east of Palma, from Coll d'en Rabassa to El Arenal, as performing this role by the end of the 1980s;
- the settlement that has grown up around a cove or series of coves including, for example, Calas de Mallorca and the expanded Cala d'Or;
- the resort developed around an existing town or village on the coast that expands under the influence of tourism, like Porto Cristo or Cala Ratjada. An extension of this type very familiar in Mallorca is the resort that is almost an extension of a nearby *inland* town such as Cala Millor, which is connected to Son Servera by a dual carriageway with some elements of a green belt in between;
- the linear resort or series of them that eventually merge, strung out along a long narrow beach with a grid plan of streets. There are many examples of this type, especially those along the bays of Pollença and Alcùdia in the north, such as Can Picafort, and Magaluf and Santa Ponça in the south.

Of course, any such classification is an oversimplification of reality. Many, if not most, of them have developed a number of these characteristics by the present day as they become more complex urban places, not simply single function tourist resorts.

Planning and development: moves towards a more managed landscape?

The history of coastal urban development in Mallorca is littered with attempts by town planning agencies to shape and fashion their appearance over the last one hundred years, and particularly over the last fifty or sixty. Chapter 10 showed how garden city and suburb planning ideas dominated in the early part of the last century. Since the 1950s there has been a growing determination of the governing powers in Mallorca to control their own environment, but their efforts have been subject to immense political and commercial pressures.

In the post-Civil War period the tourist industry was seen as a means of reviving the closed Spanish economy by encouraging inward investment and accruing foreign exchange capital. It seemed quite natural at the time to keep planning controls to a minimum. Apart from fairly rudimentary restrictions on size and location and a basic road plan and services, developers were given a fairly free hand to develop land. However, by 1963 Franco's government had identified a number of specific areas as tourist zones, including Calas de Mallorca, Las Gaviotas in Muro and Bahia Nova in Artà. The development of resorts and the building of large regular, block-like hotels was perceived quite differently then. They were seen as part of the modernisation of an island that only a few years previously had suffered from outward migration and hunger. They had a bedazzling effect on local consciousness (and especially on politicians!) that blinded many people to the damage they would soon do to the ecosystem. It was not long before a reaction set in and protests were mounted to save some of the island's more precious coastal resources such as the huge beach at Es Trenc and the calas of Mondragó. The environment and the landscape increasingly became seen as important tourism resources. Without controls the untrammelled growth model would soon degrade the very things visitors came for. Fortunately, many of these early proposals were never completed – demand fluctuated, capital ran out, promised water supplies did not materialise – but planning permission, such as it was, was not always lost, and many of these early urbanisations now have a rather ghostly appearance in the contemporary landscape – roads, lamp-standards and outline building plots abandoned in the countryside, relict features of unfulfilled ambitions. Although an Act was introduced in 1965 to control tourism in Spain, it failed to tackle serious planning issues, and overdevelopment and low grade urban landscapes became the norm in much of Mallorca. Some municipalities decided to try to implement their own controls on resort expansion. In one of the northern bays, for example, Alcúdia town council brought in its own plans in order to avoid some of the worst excesses that were taking place elsewhere on the island, but in aesthetic terms it was not very successful. Between 1966 and 1970 seventeen hotels were built, all displaying pretty functional architecture and built of the ubiquitous reinforced concrete (Vidal, 1994).

A major weakness at that time in planning for Mallorca's boom industry was that early legislation tended to be permissive rather than restrictive, and even so, it was often ignored. When a large degree of autonomy came to the Balearic Islands after 1983, planning became more of a local responsibility, but to begin with it made little difference because by then a culture of 'illegality' was well established, and any new legislation took a long time to implement. The Cladera Acts of 1984 and 1988 called for greater space allocations to be made for each tourist bed-space, rising from 30 m² to 60 m². In order to try to protect certain kinds of landscape the Areas of Special Interest Act were introduced in 1984, new coastal protection decrees came in 1988, and the first of the natural parks – Albufera – was designated in the same year. The late '80s marked a turning point for planning in Mallorca, and over the next twenty-five years much more stringent controls on building, especially in the countryside, were brought in. However, implementation was in the hands of the municipal authorities, and experience varied across the island (Buswell, 2011: Chapter 7). Most of the plans were aimed at improving the quality of the tourist experience, largely a move to improve the image of

Mallorca's holiday zones following the detrimental effects of the so-called lager-lout syndrome prevalent in many of Calvià's resorts (Buswell, 1996: 312). In terms of the TALC model mentioned above, legislation in the 1990s began attacking the degenerative phase by trying to stimulate rejuvenation to kick-start a new cycle of development. Whereas the early development of tourism was essentially a private sector activity overseen by a benign State, from now on any correction to failings in the market became the responsibility of the local 'state', the Govern and the Consell. This was especially true for infrastructural provision – roads, water supply, sewage and waste disposal. The 1990s also witnessed a much greater recognition of the role of the environment in the tourist industry, partly the result of the reaction to despoliation mentioned above, and partly a response to Spain's entry into the EU in 1986 and the adoption of its environmental policies. In reality, of course, it was mostly about taking steps to retain Mallorca's market share as other parts of Spain and the Mediterranean took off.

Apartments and second homes in the landscape

From the mid-1970s the tourism landscape of Mallorca began to take on a new slant in which the hotel so prominent in the earlier history of island holidaymaking began to give way to self- or partially-catered holidays in apartments. Palma and the south coast remained characterised by hotels, with a growing proportion of apartments, but in the north and on the east coast apartments made up 36% of the accommodation by 1986. More than 50% of this building was carried out without full and proper planning permission, indicative of the low level of control being exercised by the authorities; it was especially acute on the east coast, where nearly two-thirds of the apartments built were illegal. In the Bay of Pollença only 328 apartments met legal requirements out of a total of over 3000 (Buswell, 1996: 324; table 12. 6). By the late 1990s this phenomenon – the apartment, legal and illegal – had spread throughout the island's resorts, accompanied by the 'all-inclusive' holiday whereby a tourist need never venture beyond the hotel/apartment complex's walls, giving rise to a kind of tourist compound. There were a number of benefits for the visitor in these developments including greater informality, lower costs and a supposed greater privacy. Whereas the three star hotel – the norm in Mallorca – had to be large in order to secure economies of scale, with a large number of small apartment blocks less capital was required, giving a quicker and higher rate of return. In landscape terms the hotel was a compact, intensive user of land, but the urbanisations of apartments, bungalows and chalets were relatively extensive, contributing more to coastal sprawl (Seguí Aznar, 2001: 138–9).

The second home in some ways has its origins in these apartments. Over the last century they have taken on a variety of forms and functions. Initially they were usually large (-ish) villas located near the coast with good access to Palma. By the 1970s the apartment dominated together with small, seasonally occupied houses (chalets and bungalows), again with a coastal location. As low and medium season tourism grew, there was a move towards the acquisition and conversion of inland properties, especially poor and recently abandoned farm houses. Larger ones – the *possessions* of the sixteenth to nineteenth centuries – were often converted to hotels as part of a move towards agrotourism encouraged by the authorities as part of their diversification policy. The process of second home formation tended to add to urbanisations in the coastal areas, but also gave rise to isolated and dispersed patterns in the countryside. In many of the smaller villages, property conversion stimulated general rural development. This would have been seasonal to begin with, but some were occupied on a more permanent basis as owners came to Mallorca more frequently throughout the year. It has been estimated that Mallorca has an average floating tourist population of nearly 240,000 at any one time, the majority (55%) still to be found in hotels, but the remainder now in residential accommodation, amounting to 3. 8 million of Mallorca's 8. 6 million tourists in 2006. The final phase has been the increase in more permanent residents from elsewhere in Europe – particularly Germany, from North Africa and Latin America, and indeed from many countries across the world, especially China, since about 2000. Between 1989 and 2006 net migration to Mallorca increased by over 135,000 (Llibre Blanc, 2009:220–1). Each of these periods, and the type of development they engendered, contributed new or converted elements to the landscape: a kind of coastal sprawl, punctiform building in the countryside, and a degree of village and small town revival. Of course, this has not been due to the international immigrants alone. Mallorcans themselves have rediscovered their rural roots and bought and restored houses, sometimes as second or weekend homes that later became permanent.

Until the crisis that began in 2008 Mallorcans have benefited from rising incomes in the last twenty years, leading to a considerable increase in commuting from the interior villages and towns into jobs in Greater Palma, aided by rising car ownership, investment in roads, and in part, but only recently, by improvements in public transport. The extension of the island's motorway system north towards Port de Pollença and east to beyond Llucmajor, the dualling of the main road from Palma to Manacor, improvement to most of the remaining road network, including resurfacing many of the rural *camis*, has increased the propensity for movement considerably. This has contributed to a wider process of *counter-urbanisation* particularly evident in Calvià, Llucmajor, Campos and Marratxí municipalities. However, the spread of suburbs from Palma has been going on at the same time at the expanding city's edge, so that a much wider and distinctly built-up urbanised landscape is now evident in the south of Mallorca. Greater Palma is acquiring all the features of a megalopolis (even if on a small scale by European standards) within which tourism functions are but a part. The suburban estates of Pont d'Inca, Verge de Lluc and Marratxí, the out-of-town retailing sprawl around the city's Via Cintura, the sporting and recreational facilities, the motorways and other high speed roads, the *possession* conversions, the multi-million euro villas in the gated and secured Son Vida, the coastal spread to east and west and the growth of nearby towns are all symptomatic of rising affluence. It has not only been the towns and suburbs around Palma that have been embraced by these changes; rural areas have been swept in too, as townspeople seek more isolated dwellings in the countryside proper, a process known rather clumsily in English and Catalan as '*rururbanisation*' – the urbanisation of the countryside. If these are first or principal residences they tend to be nearer Palma, such as in Santa Eugènia or Bunyola, but if second homes then in more remote locations like Sant Joan in Es Pla. If of high density, they are to be found in areas which experienced land division in the late nineteenth and early twentieth centuries; low density homes of this kind occur where the large estates survived intact longest, so the houses are often widely spaced, large fincas. The slopes of Tramuntana are attractive to many in areas like Esporles and Puigpunyent. For the more international settlers views of the countryside or the sea are often determining factors; the Serra de Llevant in the east offers both, as do Bunyabalfur and Andratx in the west. Such variables are naturally reflected in the prices such properties can command (Binimelis Sebastián, 1998). In many of these areas planning, as a welfare objective, has clearly given way to financial pressures as local municipalities have flaunted their own legislation and permitted, or in some cases ignored, what to many observers are inappropriate developments, the price, perhaps, of rising affluence.

Around all the coasts of Mallorca, a new form of second home has arisen, namely the yacht and the cruiser. By 2008 there were over 14,000 berths for these largely foreign vessels widely dispersed along the east coast, in Calvià and Andratx, the Bay of Palma and the two northern bays. The exception is where the Tramuntana descends in steep cliffs into the sea. This pattern is very similar to that of land-based second homes, adding further to the developmental and environmental pressures. The *club nàutic* of the 1930s with its small number of sailing devotees has now given way to a large number of professionally run marinas, many with large numbers of permanent berths. While some of these occupy natural ports and harbours, or perhaps began as such, many are artificial creations constructed with massive breakwaters made from enormous boulders mined from Mallorcan quarries (Fig. 11. 4). Large areas of the soft rocks of the coast have been excavated to provide these extensive berthing facilities. These marinas are, in effect, extensions of the urban morphologies to which they are attached. Figures 11. 2 and 11. 4 have been modified in a seaward direction to include them. A good example is the creation of the marina between Cala d'Or and Porto Petro on the east coast, where the interdigitation of buildings and boats makes up one complex urban landscape with nearby car parking and a proliferation of shops, restaurants and marine services. For the authorities they represent, of course, part of the move towards a higher grade of tourism, despite the fact that few of the vessels spend much time at sea (*Consellaria de medi ambient, ordenació*

Figure 11.4 The marina at Port d'Alcúdia, 2010. This marina is typical of changes to the townscape of seaside resorts. It is totally artificial in that the mole and the breakwater have been constructed by dumping tons of boulders into the shallow sea to provide protection for the hundreds of yachts and cruisers that are anchored here, many of which rarely venture far into the Mediterranean, if at all. While supported by municipios and the Govern Balear as part of the diversification of tourism policy, local residents often protest against their construction. Source: Estop.

del territori i litoral (1997)). But they are new and distinctive additions to the Mallorcan 'landscape', clearly urban in character.

Emptying the heartland: twentieth-century changes to Mallorca's centre and mountains

The take-off of the tourist industry to sustained growth between about the mid-1950s and the 1990s was accompanied by – some would say accomplished by – emptying the centre of the island of its people and its economy, resulting in a desolate and depressing human landscape. Only since the turn of the century has there been a limited revival of parts of the Pla and the Tramuntana.

In 1950 agriculture employed about 70,000, about 40% of the workforce. By 1975 this had fallen to 28,000 (13% of the workforce) and a decade later the respective figures were 16,000 and 7–8%. Previously, analyses of population have been used to describe the decline of central and mountain rural areas. However, it has to be remembered that over this time period (1971–1990) the population of Mallorca was always expanding, and therefore it is important to be able to show which areas were gaining or losing population absolutely *and* relatively. The decaying heart of Mallorca, the north-west foothills and isolated parts of the south-west coast, declined in both senses, but the north-west coast and the Tramuntana increased their population absolutely while declining relative to Mallorca as a whole. Two other points to bear in mind are, first, that by 1991 Mallorca had a highly concentrated pattern of settlement with only about 5% of the island's population being described as dispersed, and secondly, that at the municipal level many of the changes were very small, so, for example, Fornalutx (a mountain municipality) lost only eight people over the twenty year period, while Mancor de la Val increased its population by over 7%; but this represents only sixty people. Local movements have to be put into an island-wide context of demographic change. Over relatively short time periods the dynamics created by tourism could bring significant changes. Between 1970 and 1981 most of the growth was concentrated in the Palma region, but in the next decade it was much more widespread. In most cases the rural municipalities that had a coastline, except for those fronting the Tramuntana, gained in population under the impact of tourism and the attendant urbanisation described earlier. The loss of people from the central and mountain regions is a useful signpost to the landscape changes that followed in its wake.

Behind these bald statistics lies a complex story of economic and social transformation resulting in a landscape made up of abandoned farms, neglected fields, empty schools, declining rural industries and services. In the 'natural' environment the Aleppo pine spread and the *garriga* was allowed to re-establish itself after the great land clearances of the 300 years from about 1600. This was, of course, not a process that suddenly began in the 1950s. Previous chapters pointed to the high rate of emigration from the centre and mountain municipalities in the late 1800s and again following the Civil War. Farming, whether of the extensive kind of the *cerealistas* and *arboriculturalistas* of the *gran possessions* or the more intensive small holdings of the *pagès*, was rarely a paying proposition despite the success of some commercialisation after about 1850, and particularly in the 1930s. For many, if not most, near subsistence was the rule. Something like 16,000 farmers were lost in the fifty years from 1930, and those that remained were increasingly elderly. By the early 1980s four types of farming could be identified: Mediterranean dryland farming, a combination of cereals and tree crops especially in Es Pla; subsistence farming in marginal lands or in areas of more extensive pasture farming; pasture farming based on irrigated fodder production concentrated in the Campos area and between the outskirts of Palma and St Jordi; and finally, the irrigated production of horticultural crops dependent upon demand from tourism and water supply from artesian sources. This was focused on two areas – Muro/Sa Pobla in the north and the Pla de Sant Jordi in the south. However, by 1982 more than 60% of farms were dedicated to part-time dry farming, mostly of tree crops (Salvà Tomàs, 1987) Coastal land became urbanised as farming land was lost to tourism and speculative building. Between 1960 and 1970 land classified as urban rose by 20,000 ha and by 1984 by more than 40,000 ha. Historically, Mallorca rarely had had more than 50% of its area under the plough or grass, a measure of its difficult environmental conditions, so if land was now lost to building and other urban uses, it meant that farming became geographically more concentrated on better areas.

Table 11.1 Land-use changes 1960–2010.

Land Use	Hectares 1956	%	Hectares 2000	%
Quarries & Golfcourses	1.99	0.04	2.24	0.45
Urban Land	5.66	1.14	24.81	4.99
Dry Farming with tree/crops	178.22	35.85	170.08	34.21
Dry Farming without tree/crops	90.15	18.14	82.67	16.63
Olives	13.93	2.80	12.45	2.51
Irrigated with tree/crops	2.41	0.48	2.79	0.56
Irrigated without tree/crops	20.05	4.03	19.64	3.95
Woodlands	86.81	17.46	84.28	16.95
Garriga	64.98	13.07	62.98	12.67
Wasteland	31.15	6.27	31.67	6.37
Water Surfaces	3.56	0.72	3.53	0.71
	498.90	100	497.15	100

This change was accompanied by further landscape shifts away from traditional dry land farming and more towards irrigated land, where crops could be grown in response to the demands of the tourist industry and Mallorca's urban population (see table 11. 1). Three areas benefited, none strictly in the decaying heartland – the north and south *marinas* and the *raiguer* – principally determined by the massive water supply requirements of horticulture and fodder-based dairying. As in other parts of Mediterranean Spain, the polytunnel became a new landscape feature. Geographical specialisation also began to increase; for example, the Sa Pobla area, with its deep and relatively rich soil capable of irrigation, had developed its early potato industry in the 1930s, and a cooperative with over a thousand members had been set up. Disrupted by the Civil War and the Second World War, production was not resumed until 1946, but by the 1970s the operation was put on a more commercial basis, new markets besides the traditional British one were opened up and new crops introduced, diversifying away from potatoes (42,000 tons) into onions (4000 tons), beans (3000 tons) and other vegetables. To the very distinctive landscape of large, flat open fields were added new storage facilities, freezing and dehydration plants (Berga Pico, 1985). A similar cooperative was developed around Porreres for the growing and processing of apricots.

The wine-producing areas so important to the agricultural economy in the late nineteenth century were themselves affected by phylloxera, and by the mid-1980s only 3000–4000 ha were under the vine in the traditional areas of Binissalem, Consell, Santa Eugènia and Santa Maria (2000 ha), Felanitx (1000 ha), Porreres (400 ha) and Manacor. Initially production was for local markets using local varieties, Callet, Monte Negro and Fogoneu. During this century new technologies and other varieties, accompanied by a new and more dynamic entrepreneurship and a rising demand stimulated by tourists and new residents has resulted in a considerable revival of this traditional aspect of the Mallorcan rural landscape. This is another example of specialisation that is helping to provide new economic opportunities helping to revive parts of the dying heartlands.

One landscape feature in the interior that always attracts the attention of the visitor, and often infuriates local people, is the golf course with its well-watered greens, its affluent looking clubhouse and its penumbra of new and expensive houses. The Mallorcans have a sarcastic phrase for their appearance – *una sucursal de Escocia, pero de secano!* – which might be translated as 'a bit of Scotland, but in the drought!' (Borras, 1991: 89). They too are the product of processes we will see elsewhere: the availability of cheap former agricultural land, rising incomes amongst visitors, and a desire to diversify the tourism product and season. Even in the 1980s, when the number of courses was limited, 85% of the rounds were played by foreigners and the Balearic Islands received about 18% of all Spain's golf visitors (Socias Fuster, 1989). Today, some of the courses are the location of major international tournaments and not simply the place for a tourist's diversion. As in so many other countries, however, the building of houses on the same land is usually an accompanying feature, 'suburbanising' the countryside even further and adding to the visually intrusive nature of such developments (Fig. 11. 5).

In 1993 Pere Salvà wrote that the Serra Tramuntana 'is not an area with an agricultural future' (Salvà Tomàs, 1993: 92). As in the plain, only where sufficient water supply coincides with a reasonable amount of flat land can worthwhile crops be grown; in the mountains this is limited to small pockets in the dolines or *poljes*, valley bottoms and on the lower slopes. Historically tree crops have been the norm for this zone, with the olive predominant. In 1860 over 18,000 ha were given over to it, but by the early 1990s only 6700 ha were in cultivation. However, this remains the major agricultural land-use. Many of the ancient terraces whose origins we have examined are now deteriorating quite rapidly. If it is to survive on any scale an enormous reclamation programme for this very distinctive landscape will have to be instituted. Over the last 130 years farming in the mountains has lost well over 10,000 ha and pines and scrub now cover many formerly farmed areas. The woodlands, once important as part of an integrated upland economy and landscape that involved extensive pasture, charcoal making, timber production and hunting, now exhibit only remnants of these historic features. In their place have evolved new recreational activities such as hill-walking, mountain biking, bird-watching and the traversing of the mountain roads by large numbers of tourists' coaches from the coastal resorts on their way to Lluch monastery and Deià, none of which seem to contribute to the maintenance of the actual landscape, the very thing many visitors come to see. The other major competitor for land has been urban development and new housing which, by definition, has been even less respectful of their surroundings. The 1991 Law of Natural Spaces may well offer protection to the Tramuntana, but has done little for its resource management. In 2011 over 30,000 ha of the Tramuntana's cultural landscape was granted UNESCO World Heritage status, which should ensure the area receives better publicity and should give it better protection from development, complementing conservation action taken by various levels of government in Mallorca (www.

Figure 11.5 The golf course at Son Termes (Bunyola). It is clear that such features are quite alien intrusions into the landscape of Mallorca. This is because they are well watered, even if today it is with 'grey' (recycled) water, and appear green throughout the year, especially at times when so much of the landscape appears ochre in colour. Their development has been strongly resisted by environmentalists. Source: Estop.

serradetramuntana. net, June 2011). Of course, it could make the mountains even more attractive to tourists, and thereby subject to even greater pressures.

However, certain improvements to Es Pla and the mountains have been due to the considerable increase in residence in these areas. The coast may have been the original magnet for those seeking second homes, but from the 1990s the diseconomies of such areas associated with overcrowding had begun to set in. The increasingly affluent British and German middle classes began to look for fincas and more isolated farmhouses that could be bought relatively cheaply (at least to begin with) and converted to modern standards, thanks to an expanding rural electrification programme and a gradual increase in piped water. For many, of course, the attraction was in remoteness and a minimum of services. Soon '*Se vende*', '*Zu verkaufen*' and '*For sale*' became the new headlines as the island appeared to be up for sale; in fact it was more of a *crie de coeur* than a political rallying cry (Seguí, 1998). Between 1992 and 1995 something like 380 rural fincas changed hands in Mallorca, mostly on the edges of areas already made popular by tourism; areas such as Pollença and Artà in the north, Manacor and Sant Lorenç in the east and Calvià in the south-east, 80% of the transactions involving Germans (Seguí 1998: 218, table LVII). Some areas, such as the valleys around Son Macià in Manacor, indeed became virtual German colonies – at last, Bismarck's dream of a German 'place in the sun' had been realised. The benefits to local economies were not immediately evident, as many of these acquisitions and conversions acted as second homes initially, but as permanent immigration increased, then not only builders benefited from the new settlers. Restaurants, bars, cafés and supermarkets began to notice an all-year trade developing. Immigrants also began to establish their own businesses in nearly all sectors, but especially in those associated with real estate and specialist retailing. As with the immigrations from North Africa and Latin America that followed in this century, the landscape effects were less noticeable than the cultural ones as the traditional way of life of Mallorca's heart was undermined. Mallorcans sometimes forget that the causes of these changes in the heartland were not restricted to incomers alone. Below we shall show that while much of the population expansion of the late1990s and early 2000s was accommodated in the Palma city region, there has also been a marked increase in decentralisation to more distant parts of the island, although this has been restricted to the larger towns – Felanitx, Inca, Manacor, Artà – especially where they had a coastline – and above all the towns of the raiguer. Recent population increases in the true heart of Es Pla were about 8% for 2001–2006, but the absolute increase was less than 2000 out of an island-wide increase of over 122,000. Nonetheless, new housing opportunities in the adjacent countryside and smaller villages have been growing. In addition to long distance commuting, short distance journey-to-work and to shop movements to the larger inland town have also increased, thanks to Mallorca having the second highest level of car ownership in Spain, after the Commune of Madrid.

To some Mallorcans of a certain age, the deterioration of the historic cultural landscape of Es Pla is to be regretted; to them this was the true Mallorca, *Mallorca profonda*. However, there is a danger that in an effort to regenerate the area and to provide the kind of economic opportunities that are available in the coastal resorts and Greater Palma, Es Pla might become something of a theme park based on its cultural 'preservation' with, in effect, 'museums' of former ways of life. This commodification of the countryside by a heritage industry is seen by many as a dangerous development towards falsifying or possibly perverting history for economic purposes. The idea of Es Pla as the 'real' Mallorca is probably a dangerous fiction, according to Climent Picornell. Using it as the basis for an economic geography of inland tourism could lead to mistakes similar to those found on the coast (Picornell and Picornell, 2008).

Calvià: a case study

One *municipio* that demonstrates this landscape transition from rural backwater to a highly urbanised environment is Calvià, the zone in the south-west of Mallorca that has benefited most from early expansion, but that has also suffered some of the worst pressures of uncontrolled growth. In the nineteenth century its landscape was one of figs, almonds, olives and carobs and, of course, pine trees, grown on some of the largest landed estates on the island, including that of the Marquis of Romana. In 1857 its population was only 2308, about the same as in 1960. People and work were focused on these estates, controlled by a negligent aristocracy or by the new *haute bourgeois* class living in Palma. In 1920 65% of employment was in agriculture, a figure that had fallen dramatically to 7 or 8% by 1960. In the pretouristic period most of the work was for day labourers; for example, the Santa Ponça estate employed only 36 on a permanent basis in 1857. It was hardly surprising that during the last twenty years of the nineteenth century, and again in the 1940s, Calvià suffered from emigration overseas. It was this class of poverty-stricken farm labourers that went overseas. In the 1950s,'60s and'70s it was the remainder of the same class that left the land and took up the new opportunities in nearby coastal tourism. Later the small proprietors abandoned their *estabilments* and their *parcelas* too, to supplement their incomes, becoming part-time farmers. By 1970 population had risen to 5000, but to 18,000 by 1975. Densities were 20 per square kilometre in 1960 but 123 in 1975 (Salvà Tomàs, 1983).

As we saw in the previous chapter, Palma Nova, only 14 km from the city, was developed in the'30s as a 'garden city', in effect a suburban expansion of the capital. Most of the resorts only begin in the 1950s, for instance Santa Ponça, planned to have a capacity of only 15,000 visitors. Nearer the city *palmasanos* had used some of the numerous coves for weekend houses and as places for fiesta outings. By 2003 Calvia had 42,000 residents, 6 tourist zones, 18 nucleations, 54 km of coastline, 27 beaches and 120,000 bed-spaces in more than 250 establishments. Twenty per cent of Mallorca's visitors came to Calvià, that is over one and a half million of them. Forty years before this these figures were very different. Then the municipality had only 6800 bed-spaces in 112 establishments. The municipality went from being one of the poorest places in Spain to being the country's richest (Dodds,

2007). Besides the natural benefits of the beach and sea resources, developers were able to purchase front-line sites with relative ease because the value of local land for agriculture had fallen to almost zero by the 1960s; there were few planning restrictions. More adventurous construction technologies also enabled them to build on heavily dissected rocky land right on the coast, and on steep terraces behind. Once the dense network of hotels and later apartments and second homes was established by the 1980s in an almost continuous arc that included the now well-known resorts of Palma Nova, Santa Ponça, Magaluf, Peguera, Portals Nous, Illetes and Costa d'en Blanes, then other infrastructural elements were soon added (Fig. 11. 6). The most important was the extension of the motorway west from Palma, and soon a wide range of recreational and amusement facilities ranging from golf courses to waterparks followed. All this stimulated rapid development that needed a growing workforce for construction and services. Calvià could not provide this, and the municipality's population grew swiftly, increasing by a factor of eleven between 1970 and 2000. In the early decades most of this was achieved by immigration, largely from the poor regions of the Peninsula such as Andalucía. However, by 2000 over 50% of the population was under 35 – the children of the immigrant generation – and only 7–8% over 65. By 2006 the population had risen to over 45,000, a five year increase of more than 16% achieved by a combination of overspill from Palma as Calvià becomes part of its city-region, together with a considerable influx of European settlers from the UK and Germany. These factors have led to a substantial building programme for housing, this alone shifting the landscape further and further from its rural origins to its present-day urban appearance.

Figure 11.6 Two beaches at Peguera (Calvià). Resort development is in a narrow linear band facing the sea. The beach itself is almost certainly maintained artificially. Such tourism landscapes are almost continuous around the coast of Calvià, making it one of the richest municipalities in Spain. Source: Estop.

Calvià has a high concentration of residential tourists (i. e. not in hotels), 34,000 on average *per diem* or a total of 1. 2 million in 2006, increasing the local resident population of the area by a factor of five in the high season. This is a remarkable record, but the environmental and landscape price for accommodating visitors and residents alike has been very high; to quote Miranda Gonzalez, Calvià was *'creció por el turismo y vive para el turismo. Los hoteles y las urbanizaciones caracterizan su paisaje'* (Miranda Gonzalez, 2001: 21). ('Calvià was created for tourism and lives by tourism. The hotels and the urbanisations characterise its landscape.') Perhaps the one resort that epitomises the creation of an urban landscape that is almost entirely the offspring of tourism has been the growth of Magaluf, known to thousands of British tourists, especially youth tourists (Gonzalez Perez, 2003: 146–50). Here has been created a scene that provides many of the elements of leisure found in Manchester, Newcastle and a thousand smaller British towns. A landscape of excesses: cheap three-star hotels and apartments, bars and all-day and all-night drinking establishments, usually with wide screen televisions showing Premier League football, cheap eateries offering only English cheap food (the all-day breakfast, fish and chips, the ubiquitous pizza), clubs and discos – a landscape as much of the night as of the daytime, where the beaches (artificial) offer somewhere to sleep off the night's excesses rather than then being the theatre of display envisaged by the resort's founders (Fig. 11. 7). 'Magaluf is not the creation of Mallorcan culture but that of any British city or town on a Saturday night'

Figure 11.7 Magaluf (Calvià) is perhaps the archetypal mass tourism resort in Mallorca, even notorious. Its excellent beaches and cheap three star hotels and apartments attract a young clientele for whom the night time is as important as the day. The townscape and the commercial signage have little reference to Mallorcan culture, leading some observers to describe such places as 'concentration camps' for tourists.

(Buswell, 2011: 174). Like Alcúdia, Calvià was one of the municipalities that tried to develop its own policies at a time when national efforts were minimal. This attitude was resuscitated in the 1990s under the POOT (*Plan d'ordinació de oferta turistic*) schemes and more recently under the EU's Agenda 21. The latter has been used to move tourism into a much more sustainable mode, and while there has been some success, the initiative has been weakened by problems of defining 'sustainable' and by local political corruption. Nonetheless, while sustainability may be proving to be a means of rejuvenating Calvià's tourist industry, its landscape has been irredeemably altered (Robledo and Batle, 2002).

The last decade: one geographical system but many landscapes?
The somewhat belated recognition that tourism as practised was not environmentally sustainable came at a time when the nature of the holiday experience was changing. In the late 1990s and early twenty-first century Mallorcan tourism agencies were trying to shift the emphasis from the high season and from 'sea, sun and sangria' holidays towards a broader range of recreational and leisure activities throughout the year – in other words, easing the pressures from the mass market and moving towards middle-class calendars and pursuits. This coincided with the two other processes to have a profound effect on island landscapes that we identified above: self-catering holidays in apartments and the acquisition of second homes facilitated by, secondly, the introduction of cheap, no-frills airline flights from Europe, especially from Germany and the United Kingdom. However, this process has not been as detrimental as many imagined. In the 1990s Mallorca's birth rate declined to near zero, and between 1991 and 1996 the annual average increase in population of over 12% was achieved almost solely by immigration. And the geography of this type of growth changed too. By 1996 Palma's share of population had fallen to 45% (from more than 50% a decade earlier) whereas the *part forana's* share rose to 55%. In the public sector the reopening of the railway to Manacor and Sa Pobla in 2002 and the establishment of Transports Islas Baleares (TIB) to integrate it with a vastly improved bus service now readily permits a more environmentally friendly means of transport. In Palma the construction of the new underground intermodal station has created a new node for public transport where train, bus and the new underground Metro out to the University all meet. The Antich governments of the last two decades then invested heavily in expanding the rail system to mirror the network that existed prior to the closures of the 1960s, only to have the considerable civil engineering works that were undertaken abandoned as a result of the economic decline of the period following 2008 and the defeat of PSOE at the polls in 2011. Surely, these embankments, cuttings, bridges and new stations will not now become yet more relict features to add to Mallorca's changing landscape? (See Fig. 11. 8.)

   As in other parts of the European urban system, the social, economic and spatial processes behind the creation of the Palma city-region are leading to a no-place

Figure 11.8 The rebuilding of the railway to Artá from Manacor was undertaken in the last decade as part of the plan to improve public transport. However, it remains incomplete, a victim of a change of government in the Balearic Islands and the deep recession that began in 2008 – a relict feature in the landscape even before it was finished.

geography as tourism landscapes merge seamlessly with those of a present-day, post-tourism era. In the previous two chapters the Mallorcan economy was shown to be industrialising often at a rate faster than that in Spain as a whole. This process continued well into the twentieth century, despite the tertiarisation of the economy via tourism. The successful industries of the early 1900s – footwear, textiles, food processing – have now largely given way to industries that service tourism and the growing needs of a post-touristic society. As in other parts of Europe much of the industry is less concerned with actually making things, but rather concentrates on storing, distributing and selling them, and on processing information in one way or another. The Llibre Blanc for Industry published in 1992 focused on the small workshop activities so characteristic of Mallorca's secondary sector, noting the integration on many activities such as the furniture industry in Manacor and leather in Inca, characterised by their large number of very small workshops employing fewer then five workers (Consellaria de commerc i indústria, 1992). Ten years later many of these have either been merged or more likely disappeared, especially from the town centres of the inland towns, to be replaced by much larger factories on the *poligonos* (industrial and trading estates) located on the periphery where access by road and flat land suitable for large modern industrial and distribution building are available in contrast to the cramped sites in more central locations. This process has created a new and distinctive element in the island's landscape, especially around Palma, but every inland town now seeks to have its business park or trading estate. Spatial decentralisation is paralleled by another ubiquitous movement – that of globalisation, as prevalent in Mallorca as anywhere else. This has been helped by the EU recognising Mallorca's relative geographical isolation under cohesion policies and offering subsidies on certain goods imported into the island. One of the current ironic sights in Palma and in other large towns is that of Chinese shoe shops in an island that pioneered the footwear industry. Thousands of tourists have visited the artificial pearl factories of Manacor since their foundation in the early twentieth century, but now the largest of them has disappeared to be replaced by apartment blocks, and parts of the business relocated to the local *poligono* following threats to relocate the company to China (Sansó Barceló, 2009: 304–5).

Alongside Spain as a whole, Mallorca has suffered from the world-wide downturn of the last few years that affected the tourism industry. However, for demographic reasons Mallorca is facing something of a housing shortage, and the substantial house building programme going on in the island to provide accommodation for the rapidly increasing population has been less affected than in the 'neighbouring' mainland *costas*. It does not have the landscape of abandoned half-built housing estates seen in Murcia, Valencia and parts of Andalucía. The island's population in 1996 was just short of 610,000; by 2006 it had reached in excess of 790,000 and is now over 870,000. The islands as a whole now have over 1.0m registered residents, into which over 12.0m tourists are decanted each year. In the five years 2001–2006 Palma increased its population by almost 30,000 or over 8%. The greatest absolute and relative increases have been in the municipalities surrounding Palma – Marratxí with an increase of nearly 8000 or a remarkable 33.5%, Calvià with nearly 6500 or 16.6% and Llucmajor with 6500 or 26.8%. Large increases have also been recorded in northern coastal areas: Alcúdia over 3000 (25%), Capdepera 2500 (28%) and Pollença nearly1800 (12%). Some of the rural municipalities had much higher rates, although small absolute numbers. For example, Consell's population jumped 26.7% but only added 642 people; Sencelles 23.9% and 529 people.

One of the variables that affects the changing landscape of housing is population numbers. The rising birth rate, especially among recent immigrants, is creating new demand for housing. The rate of household formation is also increasing as divorce rates rise, people live longer and the consequent demand for single-person housing increases. In every town in Mallorca flats, chalets and family houses are now being built, and until this demand is met Mallorca will continue to experience some of the highest house prices in the whole of Spain; in 2004 house prices rose by an average of 18%, only to fall somewhat after the 'crack' of 2008–12. Also, the local people have to compete with the growing demand for permanent and second homes from foreigners and immigrants who, in the case of the north Europeans, often have considerable capital at their disposal following the massive price inflation of their own house prices. The new immigrants from North Africa, Latin America and elsewhere are usually at the bottom of this hierarchy and are taking rented properties no longer required by Mallorcans, usually in the least favourable sections of the island's towns. However, the process of gentrification in Palma is taking over and renovating large areas of the inner city such as Calatrava and Puig St Pere, which a generation ago would have provided poor quality but low cost apartment housing in ancient buildings. With the aid of government subsidies for urban conservation, the middle and professional classes are taking over many of the town houses of the former gentry that from the early twentieth century lapsed into multi-occupation. In addition, public bodies such as Mallorca's many layers of government and their bureaucracies are also converting large historic properties in Palma into offices. Another competitor is the boutique hotel. The effect of this gentrification is to push the newer immigrants into the *eixample* and beyond, often distant from their places of work. One new feature of the townscape is the mosque, a feature that has not been seen in Mallorca for almost 800 years. Perhaps they might remind Mallorcans of part of their cultural and landscape heritage.

The landscape today

By the early years of this millennium the landscape of much of Mallorca had been radically altered. In social and economic terms the tipping point had been reached at which it would no longer be possible to reverse or even modify substantially the spatial organisation of the island. Its transformation from a Mediterranean destination that once attracted honeymoon couples and the more adventurous package tourists, into one of the most developed parts of the Spanish state and the European Community, had taken less than sixty years. Certain locational and environmental processes have been set in train that will now continue to shape the island's appearance for the proximate future, especially as more seasonal visitors become translated into permanent residents. The visual contrast between coast and countryside may well remain, but the urban system now extends throughout the island and true 'rurality' increasingly seems confined to only certain parts of the island's landscape, perhaps to part of a romantic past. Whilst a few years ago it might have been possible to see certain aspects of the Mallorcan countryside in their true colours in the low season when most of the visitors had gone home to Düsseldorf, Manchester and Amsterdam, as the tourist curve flattens out across the year the landscapes of the seasons begin to merge.

12
_____

Reflections on a theme of landscape change

This book has shown how the material landscape of the Mediterranean island of Mallorca has developed over four and a half thousand years: how the processes of change – economic, social and political – have brought about new and shifting geographical and perceived patterns on the ground. The dynamics may have been slow and barely perceptible over many of those centuries, but at other times much more rapid changes have taken place, such as after the invasion and conquest of 1229, at the end of the *ancien régime* in the first half of the nineteenth century, and especially under the impact of mass tourism from the mid-1950s. Prehistoric change was slow and very gradual, the Roman imprint very small, leaving few reminders of its presence. The agricultural landscapes of the late medieval and early modern eras certainly introduced new elements, but over a very long historical period. What we have also shown is that there has been a considerable persistence of all these landscape phases, one upon, and after, another; all have left their mark. The palimpsest has survived the re-writing of its surface by the dramatic transformations of tourism, urban development and new transport systems of the last seventy years. What is also evident is that much of the landscape is under considerable pressure, and that public opinion is concerned about the scale and rate of change. Of course, neither the relative scale nor the rate are new phenomena, but perhaps the present generation is one of the first to register concern and even opposition to such changes in Mallorca's landscape. Now, there is an apparently widespread concern for 'loss'.

Whose landscape?

This study set out to deal with the material landscape, the landscape of the everyday (Jackson, 1984), interpreted as the changing cultural imprint on a physical surface, itself subject to change. It sought to identify the principal historical periods through which landscape formation and construction has passed and adopted the traditional palimpsest metaphor to write in this way. It dealt primarily with what we could experience through the soles of our boots/shoes and with our eyes, but hardly at all with perception via our other senses of smell and touch. We also said that the interpretation of Mallorca's landscape would be that of a white, middle-class occasional visitor and part-time sojourner of a certain age.

This raises the question of 'ownership' of this landscape. Writing this book has inevitably been a discovery, almost literally an exploration, but having to hand only the tools inherited from a largely British tradition of techniques of landscape history. On completing most of the text one is confronted with the realisation that, of course, this is someone else's landscape; your author has little proprietary claim to it, despite having a house on the island. I have been the outsider looking in, no matter how much 'research' has been undertaken. It raises the question then: 'to whom does this landscape belong'?

One thing our chronological approach reveals is that Mallorca has long been a contested space. The early landscape was a scene of conflict between, for example, the early prehistoric settlers who sought to exploit its meagre resources for their Bronze Age culture, 'fighting nature' to defeat the *myotragus balearicus* for a food supply, and 'mining' rocks and rough timbers for shelter. We have shown that this struggle between social organisations and nature in what is in fact quite a harsh environment has been a constant theme, including present-day struggles with water supply and the management of extreme events. Intervening in this contest has always been technology, itself a creator of landscape. Within this study there are many examples of the effect of technology as an agent of change, mediating between people and their environments. I have written more than once, for example, that the aircraft 'invented' tourism in Mallorca and hence the landscape it engendered (Buswell, 1996 and 2011). A much earlier example would be the hydrological techniques of the Arab and Berber settlers that were introduced in the tenth and eleventh centuries and formed the basis for water management to at least the mid-nineteenth century. At another level the landscape is clearly the product of political forces, whether they were rooted in the feudal superiority of the medieval era with a landscape created in the lord's image, or in the redistributed lands post-Mendàzibel reforms in the nineteenth century.

Landscape values are attached to the cultural and technological values of the various 'peoples' who have inhabited the island; waves of immigrants, sometimes conquerors, sometimes peaceful settlers, nowadays rapacious temporary and transient tourists. And it was the tourism-powered immigrations from the mid-1950s that saw perhaps the greatest potential change in the perceptions and evaluations of Mallorca's landscape since the Christian invasion of the thirteenth century. The successive booms of that industry have brought in Peninsular Spanish, most of whom not only settled in Palma's poorer inner suburbs and in the new coastal urbanisations, but actually built them. They have been succeeded in more recent years by settlers from Germany (now 33,000), Morocco (20,000) the United Kingdom (10,000), and various countries in Latin America that together total nearly 40,000. Each in their different way adds something to the landscape – there's even a cricket ground near Magaluf – and they see and use the landscape in perceptibly different ways. In addition, European immigrants are also bringing in inward investment in areas like retailing and house building, so that styles and signage initially seem alien (Andrews, 2011: Chapter 3). Increasing globalisation will only add to these accretions. Will they be at the expense of the indigenous cultural norms, or in time become part of a new Mallorca scene?

Clearly this reveals another underlying theme – that of colonisation. At times this might be described as economic colonisation, such as the early prehistoric settlement that sought to exploit the natural resources, or the twentieth century colonisation by tourism, which in its exploitation of the island's coastal resources has created 'alien' landscapes of hotels and apartments and their associated infrastructures (López-Bravo Palamino, 2003). In addition there have been the more obvious imperial colonisations of the Romans and the Catalan/Aragonese, which sought political control and established new cultural hegemonies. Similarly the initial success of the Central European Habsburgs in the sixteenth and seventeenth centuries had seriously deteriorated by 1700, to be succeeded by a new French Bourbon regime that created a new order of nobility intent on making new demands on land and landscape only slightly more bourgeois than the continuing feudal system. It also brought to Palma the urban palaces that colonised medieval urban spaces.

Landscape values and the proximate future: is planning the answer?
But what of the future, in effect the *proximate* future, because it would be unwise, on the basis of the evidence we have provided, to go beyond examining the impact of processes already set in train?

What might condition any answer to this question is something that underlies both of the previous ones: do Mallorcans appreciate the *historicism* of their landscape through a detailed understanding of its formation? Are they aware of the processes that have been at work in history and geography to produce the patterns we see around us? How rooted in Mallorcan psyche is a knowledge and concern for the past? This surely is a question for education at all levels. While it is true that Mallorcan schools, colleges and the university now include environmental considerations in their curricula, could this be extended to include landscape interpretation and appreciation? Could space be found in the history and geography courses for some element of this? Certainly elements are already included perhaps as part of an unwritten attempt to'mallorcanise' what is now a diverse population.

If the landscape is the outcome of many generations of landscape change and if, in general, it is valued, then are there elements that ought to be *conserved* to enable present and future generations to understand how what they see has come into being? Clearly, we can point to examples already achieved, such as the restoration of Raixa and Alfàbia, the *possessió* of Els Calderes and the historic core of Palma with its intensive and active programme for the conservation of old buildings, including finding new uses for many of them, ranging from government offices to boutique hotels. The question one is forced to ask about these programmes is, what is the motivation? Is it to conserve such elements for their intrinsic historic value or as adjuncts to the tourism trade? Mallorca now encourages cultural tourism in its many guises, including visits to historic and cultural sites as part of its policy of diversification (Buswell, 2011: Chapter 9). This may have considerable appeal for middle-class, low-season tourists but less attraction for the mass of tourists of the summer months. If a policy of conservation is to have more success than it is enjoying at present, there is a danger of its efforts being eventually subverted by tourist numbers and true historicism slipping into the clutches of the heritage industry. The lessons of Venice, Bruges, Florence and the English Lake District are only too evident. For Mallorca individual elements of the landscape such as talayots, castles, churches and country houses need to be set into their proper contexts of history and geography, showing how they form part of wider landscapes. It is the *holistic* nature of landscapes that needs to be grasped – something that DOT (Consellaria de medi ambient, ordenació del territori i litoral, 1997) did begin to address when commenting that these *grandes unidades paisajísticas* (large landscape units) were always in danger of being fragmented, that the quality of the landscape (including its historical basis) is part of the territorial integrity (Consellaria de medi ambient, ordenació del territori i litoral, 1997: 128). In their study of Es Pla the Picornells, *père et fils*, point to the danger of misinterpreting the countryside of the centre of Mallorca as its true heart, *Mallorca profunda*, the *real* Mallorca, with the attendant danger of it then being exploited for tourist purposes so that the population of this area could receive at least some of the benefits from tourism enjoyed by the coast (Picornell and Picornell, 2008: 37). The spread of second homes and even rural hotels converted from seventeenth and eighteenth century *possessions* is already adding new, somewhat alien, urban elements to this kind of area.

But the demands of the overseas tourist are not the only concern here. Tourism is a push/pull activity, and Mallorcan ingenuity is adept at creating 'heritage' environments to attract visitors, such places having few roots in true historic identities. Just as 'the beach' has become a theatre for the acting out of fantasies and dreams, often related to the body and to sexual imagery (Baerenholdt *et al.*, 2004: 49–51; Andrews, 2011: Chapter 5), so other landscapes now operate their own stage-plays of 'Moors v. Christians', pirates and corsairs; even Easter Passion plays originally designed as part of local rituals and displays are now increasingly for the benefit of visitors. The markets in every town also have a strong theatrical element. Originally medieval foundations forming an essential part of the townscape with a true economic function that once bound the produce of the surrounding countryside to local consumption, street markets have now become retail outlets selling goods from around the world – Chinese shoes and handbags, Canarian and Azorian table cloths and embroidery, African handicrafts, Peruvian clothing. They are magnets made by Mallorcans for the tourists to flock to in their air-conditioned coaches: visit any inland town on its market day, a day probably fixed in the thirteenth century, and their popularity will be evident. Every *poblat's* fiesta, once a celebration of the local saint's day played out in its streets, is now thronged with visitors from the coastal resorts. Perhaps it is only in the low season that the landscape of rural Mallorca returns to its quiet, dignified origins.

Everywhere there are examples of historical functions being subverted for tourist consumption with little or no reference to their origins or true meaning.

It is not surprising to find that, at the governmental level, most concern for landscape and environment is with the physical and ecological impact of developments over the last sixty years, and more especially over the acceleration of development that has occurred during the last twenty, under the influence of both the tourist, and increasingly, post-touristic economies. The impact of demographic increases brought about by inward migration and the aspirational demands of earlier populations has resulted in two different spatial patterns. On the one hand much of the population growth has been contained in the network of towns and urbanisations established in the 1960 to 1980s; recently this has mostly been the effect of immigrants from Africa, Eastern Europe and Latin America. On the other hand, better off Mallorcans and north and west European immigrants have spread into the outer suburbs of the *corona metropolitana* of Greater Palma and into the island's countryside, complemented by the stability in the *pueblos* that is supported by commuting to work elsewhere. These are hardly phenomena unique to Mallorca. These processes of counter-urbanisation, rururbanisation and car-borne commuting over longer distances may be observed over much of Western Europe, not least in mainland Spain itself. The landscape being produced is one of increasing densities of building in outer Palma and its adjuncts, Llucmajor, Marratxí and Calvià, with new housing estates and apartment blocks built by large construction companies, expanding and new *poligonos*, and a rapidly growing high speed road system. The considerable improvement in the island's more distant and minor road network has made possible such dispersion of people and new economic activities into the countryside and the more accessible inland towns. Areas such as the *raiguer* – the line between Palma and Port de Pollença along the motorway Ma13 and the route east towards Santanyí and north-east towards Manacor, aided by the motorway Ma19 past Llucmajor and the duelling of the carriageway of the Ma15 to Manacor, are all axes that are guiding these changes. The on-going programme of building bypasses around the inland towns, such as around Felanitx and Son Servera, and the upgrading of more minor roads between the towns is making the central area also more accessible. The isolation of the Mallorcan countryside, once so characteristic of the landscape, is rapidly being broken. In addition to the road network and the building of houses and apartments in and around these towns is the rash of business parks and small industrial estates, providing new opportunities for work. Unfortunately the rate of occupation of many of these spaces has been somewhat slower than their developers expected, giving a rather ghostly appearance to many of them. Nonetheless every *municipio* wants one, just as they wanted to host a coastal resort in the 1960s. In the south of the island around Palma the story is different in scale and level of activity. Here the huge *poligonos* are full of factories, warehouses and distribution centres and now, many decentralised offices from

Palma itself, all accompanied nearby by the shopping malls and large scale retail activities dominated by multinational firms such as Decathlon (sports equipment, French), Leroy Merlin (D. I. Y, French), Lidl (supermarkets, German) and Eroski (Basque) and Carrefour (French) (supermarkets).

For the island as a whole two questions remain. The first poses something fundamental that places landscape on the political and policy-making agenda. How does Mallorcan society value its landscape? Is it content with what it sees around it? Has it accepted the transformations wrought by tourism and urbanisation such as those we have described, trading the benefits against the costs? Much early opinion about the post-Civil War effects was often opposed to the impact of the 'visitors', but has this now waned with the higher standard of living preferred to the environmental losses? Has the level of diversified modernity that characterises the touristed landscape now made it 'acceptable'? It would be impossible to take away the tourism industry; it is too deeply embedded in the economy and the psyche of Mallorca for that. Tourism, after all, is what defines modern Mallorca. But might it be better managed in order to control the rate of resource use and depletion, to make more effective planning measures on the one hand to prevent excesses, and on the other to produce an environmentally and an aesthetically acceptable landscape?

What has also been emerging over the last fifty years is a two-fold division in perceptions of the landscape from two different groups. The notion of dichotomy in Mallorca's geography is not a new one (Rullan Salamanca, 2002) – lowlands and mountains, coastlands and interior, Palma and *part forana*, nobility's lands and peasants' holdings, etc. – but there is clearly now a division between the landscape of the residents, which is little known by visitors, and that of the tourists in what Joan Carles Cirer has described as 'ghettos' or even 'concentration camps' (the resorts) that are rarely visited by locals unless they happen to work there (Cirer Costa, 2009: 324). Between these extremes is a less tangible source of tension between the new residential tourists and second home owners in the rural areas and the long-resident locals. Even here a third element intrudes, namely the decentralising Mallorcan locals seeking new homes in the villages and countryside and commuting to work in Palma and the larger towns.

Such dichotomies throw up tensions between 'insiders' and 'outsiders' for whom 'the past' (or for that matter, 'the present') has different significance at different times. In her study of Deià, the Mallorcan village made famous by being the home of Robert Graves, Jacqueline Waldren noted that 'outsiders maintained the material symbols of the past while the locals were struggling to find work and have enough to eat ... putting work, time and energy into improving and modernising their lives and home (while) foreigners were buying up the past as enthusiastically as the Deianencs were putting it aside' (Waldren, 1996: xviii). Later, of course, insiders/locals re-evaluated these artefacts from the past in the light of their new-found affluence and increasingly venerated them too. In the 1950s, '60s and '70s there seemed to be little opposition to the 'modernisation' of Mallorca's landscape, but from the 1990s there has been well organised, vocal opposition to so much change. This is no better witnessed than in the opposition to the road – and indeed in some parts, railway – building programmes ('Autovia, Non!'; 'Els trens: no pasarán!'). In the recent past there has also been opposition to the construction of a motorway from Palma to Manacor and to the expansion of the airport. Much of this reversal of opinion is the product of the forging of a new Mallorcan identity: a sense of isolation, backwardness and provinciality was succeeded by a modernisation wrought by tourism, followed by a re-evaluation of the island's 'past' and its true history. The tourism industry has, either inadvertently or in many cases deliberately, engendered elements that have helped create new and alien landscapes that serve visitor/outsiders and not locals. At one level this would have to embrace the whole of the urbanised, coastal, touristed landscape but at another, in the mountains or Es Pla, the artefacts are much smaller and less intrusive, but alien nonetheless (González Pérez, 2003). Tourism has, in turn, helped forge this new political identity among Mallorcans, emanating from post-autonomia liberties on the one hand, and on the other, recognition that change/modernisation had gone too far in isolating the society from its roots. The resolution of these various tensions, of course, resides with the 'realpolitik' of the island's governance. This is well illustrated with the change of government following the 2011 elections in which Parti Popular (right of centre) succeeded PSOE (left of centre) and soon announced plans to re-examine the case for increasing tourism capacity, housing and road building tempered, of course, by the exigencies of the current recession in Spain, and all likely to be strongly opposed by a variety of environmental pressure groups such as GOB and *Salvem els Paisatges de les Illes Balears*.

Planning is, of course, a political system based on a particular set of ideological values. In very simplistic terms, parties of the Right (though not the extreme Right) tend to be more in favour of market-driven solutions, whereas parties of the Left would generally support such activity.

The earliest pieces of legislation to give protection to the landscape of parts of the island were those that dealt with the protection of 'nature'; that is, to try to prevent to further erosion of the coastal lands from the depredations of the spread of urbanisation. The land-use laws of 1975 and 1990 defined the urban areas of Mallorca, restricting development in the non-urbanised parts, but its implementation was in the hands of the municipalities, who were less than stringent in its application. What followed was a wide range of planning policies emanating from the Govern Balear, ostensibly to protect the natural environment, but in fact designed to restrict the spread of urbanised land. A law of 1984 identified a dozen areas of special protection (Áreas Naturales de Especial Interés or ANEI). A 1991 law set up 49 ENPs (Espais Naturals Protegits) covering more than 37% of the island, aimed at conserving 'natural spaces', woodland and flora and fauna. These types of areas were not valued for their historical significance but for their ecological fragility (Govern Balear, 1987). It is, of course, the coast of Mallorca that has come under the greatest pressure in the last sixty years, and until the mid-1980s it was the landscapes of this zone that suffered the greatest ravishes, largely because of tourism developers' perception that sites nearest the sea ('front line') were the most valuable. The Law of the Coasts of 1988 restricted development within 100 metres of the coastline, but like so much legislation, this was only 'honoured in the breach'. It took a national law in 2009 (Article 118) to enforce this much more strictly. For the historian of the landscape the coast holds fewer challenges, largely because of the limited cultural imprint on so much of it before the 1930s. The Natural Parks that began in 1988 with the Albufera followed by Mondragó in1992, Sa Dragonera two years later and Parc Natural de Llevant in 2007 were given some historical validity, but were really seen as areas of restricted development. The designation of the Serra de Tramuntana as a World Heritage Site in 2011 was partly justified on its cultural–historical value as well as its physical geography (www. serradetramuntana. net). There are no National Parks on the British or American models, with the exception of the island of Cabrera.

For individual buildings (including gardens and monuments), archaeological and geological sites, there is a series of registers (*catalogues*) which record architectural details and historical significance, but the legislation that should assist their protection has proved difficult to enforce. Such registers are arranged hierarchically with the national *Registro General de Bienes de Interés Cultural y el Inventario General de Bienes Muebles* (BIC catalogue – *Bienes de Interés Cultural*) at the head (http://www.mcu.es/patrimonio/CE/BienCulturales/Definicion). For historic areas of towns and villages there is no equivalent to British Conservation Area policy, but municipalities can implement controls using town planning legislation such as in the Casc Antic in Palma. Each municipality maintains a catalogue of its patrimony, but lists are not a substitute for determined action. In addition, of course, the 'authorities', and especially the municipalities, might set in motion a series of reforms via new public investment criteria that are aimed at improving the worst excesses of the tourism industry. One example well under way is the pedestrianisation of many townscapes in the resorts, and not least in the historic core of Palma. The recent designation of much of the Tramuntana as a World Heritage Site clearly demonstrates the island's commitment to its mountain landscape. Despite the current (2011–2) financial difficulties faced by Spain and Mallorca's public sector, the momentum of the tourism industry and the continuing dynamism of the non-tourism economy will drive further innovations.

While the proportion of total land area given over so far to urban expansion is limited – built up and urbanised land is only about 5–6% of the total (compared with 11–12% in UK), it is highly concentrated in Greater Palma, along the *raiguer* and in the atoll of largely tourist urbanisations around the coast, except to the north and west. It was these areas that the DOT identified in 1997 as landscape areas under the greatest pressure (Consellaria de medi ambient, ordenació del territori i litoral, 1997: 111–3). In drawing up its recommendations DOT saw landscape – which in Spanish *paisaje* and Catalan *paisatge* is often used confusingly as meaning *countryside* or even *scenery* – as consisting of two sets of elements: the totality of what can be seen (the visual appearance) and a series of objects (some historical, some environmental) set in this visual context. A survey that asked respondents to rank the most highly valued areas of the island saw, unsurprisingly, Serra de Tramuntana clearly at the top, Es Pla second and the hills of Artà third; the lowest ranked areas included the corridor between Palma and Alcúdia and suburban areas around Palma and the other major towns. In another survey that asked respondents to list the elements that had contributed most to the despoliation of the 'landscape' (see above) quarries, electrical equipment (powerlines, generating plants and substations – and now photovoltaics (Fig. 12. 1), large-scale advertising hoardings and the dispersion of the industry into the countryside, along with – perhaps inevitably – the coastal tourism urbanisations, were ranked highest (worst). Surprisingly, golf courses ranked below marinas. As might be expected, valuations varied according to occupation, so the 'ecologists' ranked things higher (worse) than members of the business community (Consellaria de medi ambient, ordenació del territori i litoral, 1997: 108–10). But this refers largely to appearance. What of the cultural landscape, perhaps best expressed in Spanish as *el patrimonio cultural*? Here, in theory at least, recognition and protection is given via a series of registers of ethnographic and archaeological features that include '*naturaleza y el paisaje, muebles e inmuebles, costumbres y creencias que no son incompatibles con la vida moderna*' (natural features and landscape, buildings and real estate, customs and beliefs that are not compatible with modern life). While these recognise the richness of Mallorca's landscape characteristics, the accompanying legislation seems aimed at protecting or conserving specific artefacts, and less with the totality of landscape that would emphasise the almost ecological interdependence of all such features in an historical context in geographical spaces. Perhaps the greatest of all the pressures on landscape derives from the rising number of people on the islands – visitors and indigenous – that is associated with what might be called *late-urbanisation* (cf. late-capitalism). The demographics have been analysed and described elsewhere, with Mallorca's population now estimated at about one million, onto which must be grafted a visitor population of over nine million, the latter spread over the year but with a marked concentration in July and August (Buswell, 2011: 97–101). While the concentration of this population is to be found in Greater Palma (see above) and in the coastal tourism settlements, the effects of late-urbanisation are increasingly ubiquitous, fuelled by such mechanisms as rising housing demands quantitatively and qualitatively, counter-urbanisation, residential tourism, second home ownership, longer distance commuting, new industrial/business locations, agricultural decline and changing coastal and countryside leisure patterns. As these effects spread, so the old order of the landscape will also change, both physically and in terms of the value(s) placed on it. Policy-making and planning will play its

Figure 12.1 Photovoltaic arrays at Can Xic (Binissalem): part of the landscape of the future. Agriculture no longer pays for many farmers, so they turn to other crops! Source: Estop.

democratic part, but whether the landscape's history *qua* history can survive will depend on how the people of Mallorca and its visitors understand and value its true make-up.

Notes

Chapter 1

1 This £5.5 m research programme was chaired by Professor Stephen Daniels of Nottingham University. Its aim was to 'investigate human relationships with the natural and built, rural and urban environments, and their construction and representation. The programme will seek to gain a deeper and critical understanding of the dialogue between the environment and human agency through an exploration of the changing patterns of meaning, value and power that emerge in histories that are both material and discursive'. www.ahrc.ac. uk/landscape. See also www.landscape.ac.uk for details of research grants and awards.

2 For example, see the early part of Gomez Mendoza, J. and Mata Olmo, R. (2006): Paisajes forestales españoles y sostenibilidad: tópicos y realites *Areas*, 25, 13–30. In Mallorca the adult education section of the University mounted a series of lectures on landscape in the autumn of 2007.

Chapter 3

1 As a curious aside, Gilbert included an outline map of Mallorca in his *ex libris* bookplate, a sign perhaps of his affection for the island. He also edited the British Naval Intelligence volume on Spain and Portugal during the Second World War.

2 As a midshipman Markham 'used to read in his hammock, a book in one hand and candle in the other, and while so reading, at the age of eighty-one, under an unlit electric bulb in his house in Eccleston Square, the candle dropped, the bed took fire and he died from his injuries'. Mill, H. R. (1930): *The record of the Royal Geographical Society, 1830–1939*. London. R. G. S., 168. He had connections with Spain, having written a report on irrigation in 1867 (see Markham, C. (1867): *Report on the Irrigation of Eastern Spain*. London. Printed by Order of the Secretary of State for India, in Council.

Chapter 5

1 Some have seen Sertorius as a 'Spanish' hero who attempted to found a 'Spanish' republic. He was certainly a gifted soldier, utilising guerrilla tactics in Spain's mountains to defeat Metellus a number of times. He was finally assassinated by Perpena and Antonius, and Pompey was eventually given command over 'Spain'. The number of troops in Mallorca varies according to source; many quote 3000 as the number, but this may refer to the number of original settlers.

Chapter 6

1 Lord, 1979. The original written in the second half of the twelfth century. He owned an estate in the Gualdalquivir.

2 Arab navies had attacked the Balearic islands in 798. They pleaded with Charlemagne to defend them. Abulafia, D, 2011: 349.

3 Collins, 2004: 409–711. Collins' figures are not accepted by everyone. Pierre Guichard suggests 150–200,000 Muslim combatants entered Spain; Jean Bosch 40,000 Arabs plus 350,000 Berbers.

4 Considerable Muslim artefacts have been found in the square part of the talayot of S'Hospitalet in Manacor, for example, and amongst the ruins of the Roman town of Pollentia.

5 In Mallorca most of these qanats were quite short. For the most part they were made by local clan groups; only the larger, deeper ones were made by specialists from the mainland. See Barceló *et al.,* 1986.

6 'Place names are not set in stone. The names of tiny details in the landscape... are all subject to change. They mutate over time according to the whims of landowners, the arrival of new members into the community, alterations in the landscape, changing fashions of pronunciation and spelling, local legislation. At any one time, different communities may have different ways of referring to the same landmark.' (Hewitt, R., 2010: 161). The classic work on Mallorcan place names and their mapping is Joseph Mascaró Pasarius's six volumes published between 1962 and 1967, together with his Mapa General de Mallorca, 1958.

7 This avoidance of alcohol is itself something of a myth. There are numerous accounts of its consumption throughout the Arab/Berber period. See Unwin, 1996: 150–5.

8 The major works on the principal Arab city of Mallorca are Riera Frau, 1993 and Barceló Crespí and Rosselló Bordoy, 2006. A third, well illustrated account, can be found in García-Delgado Segués and Garau Alemany, 2000.

9 Because it frequently flooded, this river remained an important but dangerous dividing line in the city's topography until it was diverted around the Renaissance walls in 1613. See Grimalt Gelabert, 1989.

Chapter 7

1 I am indebted to the authors of an exhibition catalogue for details on this section – Jaume II i les ordinacions de l'any 1300. *Cataleg de l'exposicio,* (2002) Palma, Consell de Mallorca, Department de Cultura; especially 145–192.

2 For an account of individual churches, see Barceló Crespí, M. and Rosselló Bordoy, 2006, and Valero i Marti, G., 1992.

Chapter 8

1 Among authors see: Kamen (4 references), Elliott (3) and Casey (2)(see bibliography). On many of the maps in books like these the Balearic Islands are often omitted altogether.

2 In Mallorca in the sixteenth century a third of the wealth of the island was owned by persons living in Palma; in Manacor town 47% of the municipality's landed wealth was owned by the citizenry (see Rosselló Vaquer and Vaquer Bennàssar, 1991).

Chapter 9

1 Clearly, Elwell (1995: Chapter 3) undertook some considerable research into the Bateman family and their life in Mallorca, to which we are indebted. Very usefully he had translated a significant article in the Sa Pobla journal *Sa Marjal,* July 1916; this is included in his book, pp 36–38. For a more detailed account of Bateman's life and his work in the Albufera, see Picornell and Ginard, 1995.

2 On his father's side he was cousin of Franz Josef of Austria. He was born in the Pitti Palace in Florence, studied philosophy and natural sciences in Vienna and Prague, and spoke at least nine languages, including Catalan and Castilian. He eventually owned land and large houses in Zindis near Trieste, Ranleh near Alexandria in Egypt and Nice, as well as acquiring large areas in Mallorca (Cañellas Serrano, 1997).

3 Not that anyone had any faith in the 'English model' of landscape care by the end of the nineteenth century. Santiago Rusiñol wrote, 'When we see a place of natural beauty but

uncared for we usually say, "If only that belonged to the English. " It would be better to say, "If only that belonged to the Archduke. " A beautiful spot belonging to the English usually means the construction of funiculars, hotels and places of entertainment which destroy all its beauty; but belonging to the Archduke meant caring for the works of nature with all the love of culture and civilisation.' (Rusiñol, 1889: 176.)

Chapter 10
1 The Grand Hotel – like many of the early developments – was promoted by a Mallorcan who had made a fortune elsewhere, in this case Joan Palmer Miralles – whose money came from the footwear industry in Uruguay. It was a hotel with 100+ bedrooms and grand public rooms. It closed its doors in 1941 following the effects of the Civil War in Spain and WW II in Europe. Converted in 1942 to offices for the National Institute of Welfare, then boarded up for 40 years as a sad relic. It was restored to much of its former glory in the 1990s by Fundació la Caixa.
2 Critics said it was neither a 'city' nor a 'garden'. Many of the summer homes nearest Palma have been converted to high value permanent residences. There was always a danger of insufficient infrastructure in terms of shops, restaurants, bars, etc. Noisy flights to and from Sant Joan airport now smother the area with noise pollution every 30 seconds. The bathing facilities have disappeared entirely, as have many of the initial houses. Although the hotel was scheduled for conservation, it was in effect abandoned *en un estado deplorable*. Now new houses and apartments are being built as straightforward suburban dwellings similar to many elsewhere in Palma's suburbs.

Chapter 11
1 Parts of this chapter draw on the author's previous work. See Buswell, 1996: 309–340 and more recently, Buswell, 2011.
2 For a lucid summary of some of these economic changes see Manera, 2003.
3 For a useful account of these developmental processes, see Williams, S. (1998): *Tourism geography*. London, Routledge. Chapter 4 and Papatheodorou, A. (2004): Exploring the evolution of tourism resorts. *Annals of Tourism Research*. 31 (1) 219–237.
4 'Who shall describe adequately the paradisal sparkling freshness of its dawns, the superb pageantry of its sunsets, when the supreme Artist uses colours unknown to an earthly palette? The reds, the greens, the blues, the purples, the deep greys and blacks of the evening sky and landscape blend in such a scene of surpassing loveliness…' Goldring, 1946: 173.

References

Aalen, F. H. A. (2001) 'Landscape development and change.' In: Green, B. and Vos, W (eds) *Threatened landscapes*. London. Spon, 3–20.

Aalen, F., Whelan, K. and Stout, M. (1997) *Atlas of the Irish rural landscape*. Cork. Cork University Press.

Abulafia, D. (1991) 'The problem of the Kingdom of Majorca 2: economic Identity.' *Mediterranean History Review*. 6 (1), 35–61.

Abulafia, D. (1994) *A Mediterranean Emporium: the Catalan Kingdom of Majorca*. Cambridge, Cambridge University Press.

Abulafia, D. (2003) 'What is the Mediterranean?' In: Abulafia, D. (ed.): *The Mediterranean in History*. London. Thames and Hudson.

Abulafia, D. (2005) 'Mediterraneans.' In: Harris, W. V. (ed.) *Rethinking the Mediterranean*. Oxford. Oxford University Press, 64–93.

Abulafia, D. (2011) *The Great Sea: a human history of the Mediterranean*. London. Allen Lane.

Abulafia, D., Franklin, M. J. and Rubin, M. (eds) (2002) *Church and City, 1000–1500: essays in honour of Christopher Brooke*. Cambridge. Cambridge University Press.

Aizpurua, J. and Galilea, P. (2000) 'Property rights on the forest resources.' *The economics of Institutions in the New Millennium*. Annual meeting, Tübingen, Sept. 22–24. www.isnie.org/ISNIE00/Papers/Aizpurua-Galilea.pdf. Accessed 25 September 2010.

Aizpurua, J., Alenza, J. and Galilea, P. (2000) 'Property regimes and exploitation of the forests; an economic analysis: the case of Spain.' http://dlc.dlib.Indiana.edu/archive/00000198/00/aizupuraj041000.pdf. Accessed 25 September 2010.

Alcover, J. A (2004) 'Disentangling the Balearic first settlement issues', *Endins*. 26, 143–156.

Alenyar Fuster, M. (1984) *Introducció a l'economia de les Balears*. Palma. Caixa de Balears 'Sa Nostra'.

Alenyar Fuster, M. (1989) *Demanda, oferta d'allotjament i producció turística a Balears (1970–1987)*. Palma. Institut d'Estudis Baleàrics.

Alenyar, M. (1991) *La Industria en Baleares*. Madrid. Ministerio de Industria, Comercio y Turismo.

Alomar Esteve, G. (1943) *El plan general de alineaciones y reforma*. Palma. Ajuntament de Palma.

Alomar Esteve, G. (1950) *La reforma de Palma; hacia la renovacion de una ciudad a traves de un procesó de evolution creativa*. Palma. Mossen Alcover.

Alomar Esteve, G. (1976) *Mallorca: urbanismo regional en la Edad Madia. Les ordinacions de Jaume II (1300) en el reino de Mallorca*. Barcelona. Gili.

Alzina i Mestre, J. (1984) 'Els roters i el sistema de rotes dins l'estructura agrària de la comarca d'Artà al primer terç del segle XIX.' *Estudis Balears* 14, 17–35.

Alzina i Mestre, J. (1993) *Població, terra i propietat a la comarca de Llevant de Mallorca (segles XVII/XIX–XX): municipis d'Artà, Capdepera i Son Servera*. Arta. Ajuntament d'Artà.

Alzina, J., Blanes, C. et al. (1994) *Historia de Mallorca* vol II. 2nd Edn, Palma. Moll. Chapter 15, El Segle XX: populació i economia.

Alzina, J., Crespí, S. and Sureda, J. (1984): *Els boscos de les illes balears; la problemàtica dels incendis forestals*. Palma. Caja 'Sa Nostra'.

Amengual, B. (1903) *La indústria de los forasteros*. Palma. Amengual y Muntaner.

Amengual i Batle, A. and Cau Ontiveros, M. A. (2005) 'Late antiquity in the Illes Balears.' In: Tugores Truyol, F. (Co-ordinator): *El mon rómà a les illes Baleares*. Catalogue of exhibition. Barcelona. Fondació 'la Caixa'.

Amengual i Batle, J. (2003) 'El segle V: de la romanitat politica a la cultural a les Balears.' *Mayurqa*. 29, 157–172.

References

Amengual i Quetglas, J., Cardell i Perello, J. and Morantai i Jaume, L. (2003) 'La conquesta romana i la planificació del territori a Mallorca.' *Mayurqa*. 29, 13–26.

Andreu Galmés, J. (2000) 'Les ordinacions de Jaume II de Mallorca per a la creació de viles (any 1300): planificació urbana en quadrícula i dotació de serveis. El cas de Petra.' In: Salvador Claramunt, R. (Co-ord): *El món urbà a la Corona d'Aragó del 1137 als decrets de Nova Planta : XVII Congrés d'Història de la Corona d'Aragó* (Congreso de Historia de la Corona de Aragón: Barcelona. Poblet. Lleida, 7 al 12 de desembre de 2000 : [actes]) 3, 2003, 11–28.

Andreu Galmés, J. (2002) 'L'urbanisme planificat segons les ordinacions de Jaume II.' In: *Jaume II i les ordinacions de l'any 1300*. Cataleg de l'exposició. Consell de Mallorca. Departament de Cultura. Palma.

Andrews, H. (2011) *The British on holiday: charter tourism, identity and consumption*. Bristol. Channel View Publications.

Andriotis, K. (2003) 'Coastal resort morphology: the Cretan experience.' *Tourism Recreation Research*. 28, 67–76.

Andriotis, K. (2006) 'Hosts, guests and politics: coastal resort morphology change.' *Annals of Tourism Research*. 33 (4), 1079–1098.

Androver Rosselló, P. (2002) *Alquerías y viejas 'possessions' de Manacor*. Manacor. Grafiques Muntaner.

Anelyar Fuster, M. (ed.) (1990) '30 anys de turisme a Balears.' *Estudis Balearics*. 37–38. Palma. Govern Balear.

Appleton, J. (1975) *The experience of landscape*. Chichester. Wiley.

Aramburu-Zabala Higuera, J. (1998) *El patrón de asentamiento de la cultura talayotica de Mallorca*. Palma. El Tall.

Aramburu-Zabala Higuera, J. (2002) 'Analisis comparativo del patron de asentamiento de la cultura balearica.' In: Waldren, W. H. and Ensenyat, J. A. (2002) *World Islands in Prehistory: Insular Investigations. Vth Deia International Conference of Prehistory*. BAR International Series 1095. Oxford, 517–528.

Aramburu-Zabala Higuera, J. and Riera Campins, D. (2006) *Ses Païsses (Artà); talaiots i murades*. Palma. El Tall.

Archduke Ludwig Salvador (1882) *Die balearen in wort und bild geschildert*. Nine vols. Leipzig.

Archduque Luis Salvador (1882) *De la casa, pesca y navegación: parte de la obra Las Baleares descritas por la palabra y el grabado*. Palma. Alcover. Spanish trans. 1962.

Archduque Luis Salvador (1882) *La Ciudad de Palma*. Palma. Olañeta. Spanish trans. 1954.

Arnold, T. K. (1849) *A first classical atlas intended as a companion to 'Historiae antiquae epitome'*. London. Rivingtons.

Baerenholdt, J. O., Haldrup, M., Larsel, J. and Urry, J. (2004) *Performing tourist places*. Aldershot. Ashgate.

Barceló Crespí, M. (1988) *Ciutat de Mallorca en el transit a la modernitat*. Institut de Balears. Palma. Conselleria de cultura, educacio i esports.

Barceló Crespí, M. (2002) *La talla de la Ciutat de Mallorca, 1512*. Palma. Edicions UIB.

Barceló Crespí, M. (2012) *El raval de mar de la ciutat de Mallorca (Segles XIII–XV)*. Palma. Lleonard Muntaner.

Barceló Crespí, M. and Kirchner, H. (2000) 'L'alqueria de Felanitx: un assaig de descripció.' *Journades d'Estudis locals de Felanitx*. 348–360.

Barceló Crespí, M. and Rosselló Bordoy, G. (2006) *La Ciudad de Mallorca: la vida cotidiana en una ciudad mediterránea*. Palma. Lleonard Muntaner.

Barceló Crespí, M., Ferra Martorell, R. and Severa Sitjes, B. (1997) *Les possessions de Porreres: estudi històric*. Porreres. Ajuntament de Porreres.

Barceló, M. (1996a) 'La cuestión del hidraulismo andalusí.' In: Barceló, M. et al., (1996) *El agua no dueme; fundamentos de la arqueología hidráulica andalusi*. Granada. Sierra Nevada 95, 13–47.

Barceló, M. et al. (1986) *Les aigües cercades; els qanat(s) de l'illa de Mallorca*. Palma. Institut d'Estudis Baleàrics.

Barceló, M. et al. (1996) *El agua no dueme; fundamentos de la arqueología hidráulica andalusi*. Granada. Sierra Nevada 95.

References

Barceló Pons, B. (1959) 'El desarrallo del cultivo de la vid en Mallorca.' *BCOCIN*, 624.

Barceló Pons, B. (1962) 'Evolucion de la población en los muncipios de la isla de Mallorca.' *BCOCIN*, 637.

Barceló Pons, B. (1963) 'El Terreno. Geogràfica urbana de un barrio de Palma.' *BCOCIN*. 640, 125–178.

Barceló Pons, B. (1966) 'El tourismo en Mallorca en la epoca de 1925–36.' *BCOCIN*, 651–2.

Barceló Pons, B. (1970) *Evolución reciente y estructura de la población de las Islas Baleares.* Madrid. CSIC.

Barceló Pons, B. (1972) 'El mediterráneo occidental y sus islandés.' *Mayurqa*. 7, 5–25.

Barceló Pons, B. and Frontera Pascual, G. (2000) 'Historia del turismo en las Islas Baleares.' In: Tugores Truyol, F. (Co-ord) *Welcome! Un siglo de turismo en les Islas Baleares.* Barcelona. Fundación 'La Caixa'.

Barceló Ramis, G. (1998) 'La difusió del conreu de la vinya en el terme de Llubí a finals del segle XVI i inicis del XVII.' *Jornades d'Estudis Locals a Porreres*, 5–16.

Barker, K. and Darvill, T. (eds) (1997) *Making English Landscape: changing perspectives – papers presented to Christopher Taylor.* Oxbow Monographs 93. Oxford. Oxbow Books.

Barnabe, G. (1993) *The railways and tramways of Majorca.* Brighton. Plateway Press.

Bender, B. (2002) 'Time and landscape.' *Current Archaeology.* 43, S4.

Benedictow, O. J. (2004) *The Black Death 1346–1353. The complete history.* Woodbridge. Boydell Press.

Bennàssar Vicens, B. (2001) *Procés al turisme: turisme de masses, immigració, medi ambient i marginació a Mallorca (1960–2000).* Palma. Lleonard Muntaner.

Berbiela Mingot, L. (2010) 'Estimación de los recursos forestales arbolados de las Islas Baleares a lo largo de los últimos 250 años.' In: Mayol, J. *et al.* (eds) (2010) *Homenatge a Bartomeu Barceló i Pons, geògraf.* Palma. Lleonard Muntaner, 547–56.

Beresford, M. (1967) *New towns of the Middle Ages: town plantation in England, Wales and Gascony.* London. Lutterworth Press.

Berga Pico, F. (1985) 'Cooperativo agrícola poblense.' In: *El Campo: Boletín de informativa agraria*, 100. Banco de Bilbao, 47–48.

Bernat y Roca, M. (1997) 'Feudalisme i infrastructura artesanal: de Madina Mayurqa a Ciutat de Mallorca (1230–1315).' *Bolletí de la Societat arqueològica Lul. liana*, 53, 27–70.

Bernat i Roca, M. and Serra i Barceló, J. (2001/2): 'El darrere recinte: els inicis de la quinta murada de Ciutat de Mallorca (S. XVI).' *Estudis Baleàrics*, 70/1, 37–60.

Bestard, B. (2011) *Cròniques de Palma.* Palma. Ajuntament de Palma.

Bibiloni Amengual, A. (1988) *El comerc exterior de Mallorca. Homes, mercats i products d'intercanvi 1650–1720.* Palma. El Tall.

Bibiloni Amengual, A. (1992) *Mercaders i navegants a Mallorca durant el segle XVII: l'oli com indicador del comerç mallorquí (1650–1720).* Palma. El Tall.

Bibiloni Amengual, A. and Pons Pons, J. (2000) *La indústria del calçat a Lloseta, 1900–1996: organització i força de treball en el canvi de localitat agrícola a centre industrial.* Binissalem. Di7 Edició.

Bibiloni Rotger, J. (2002) *Palma: història del tramvia elèctric.* Palma Ajuntament de Palma.

Bidwell, C. T. (1876) *The Balearic Islands.* London. Samson, Low, Marston, Searle and Rivington.

Binimelis Sebastián, J. (1998) 'Les àrees rururbanes a l'illa de Mallorca.' *Estudis Baleàrics*, 60–61, 185–203.

Binimelis Sebastián, J. (2006) 'La difusió residencial a l'espai rural de l'illa de Mallorca a la dècada dels noranta. Noves aportacions per una correcta interpretació de l'anomena't 'tercer boom' turístic.' *Revista electrònica de geografia y ciences socials. Nueva serie de Geo Crítica. Cuadernos Críticos de Geogràfía Humana.* 10, 225.

Binimelis Sebastián, J. and Gonzalez Olivares, A. (1990) 'El tràfic de vins a traves del port de Felanitx.' *Estudis d'història econòmica.* 2, 115–129.

Binimelis Sebastián, J. and Ordinas Garau, A. (2012) 'Paisatge i canvi territorial en el món rural de les Illes Balears.' *Territoris* 8, 11–28.

Bisson, J. (1969) 'Origen y decadencia de la gran propriadad en Mallorca.' *BCOCIN*, 665, 161–188.

Bisson, J. (1977) *La Terre et l'homme aux iles Balears*. Aix-en-Provence. Edisud.

Bisson, T. (1986) *The medieval crown of Aragon*. Oxford. Clarendon Press

Blanes, C. *et al.* (1990) *Les illes a les fonts clàssiques*. Palma. Font.

Bolos y Capdevila, O. (1975) 'Paisaje y ciencia geográfica.' *Estudios Geográficos*, 36, 138–139, 93–105.

Bonzami, R. M. (2000) 'Territorial boundaries, buffer zones and sociopolitical complexity: a case study of the nuraghi on Sardinia.' In: Tykot. R. H. and Andrews, T. K. (eds) *Sardinia in the Mediterranean: a footprint in the sea*. Sheffield. Sheffield Academic Press, 210–220.

Boomert, A. and Bright, A. (2007) 'Island Archaeology: In Search of a New Horizon.' *Island Studies Journal*, 2, (1), 3–26.

Borras, M. A. (1991) 'Agroturisme i turisme rural: camps de golf i ports esportius.' In: INESE: *Turisme i medi ambient a les illes Balears*. Palma. El Tall.

Boulanger, P. (1983) 'Marsella i el comerç dels olis de las Balears al segle XVIII', *Jornades d'Estudis Històrics Locals IV*. Palma.

Bover, P. and Antoni Alcover, J. (2003) 'Understanding Late Quaternary extinctions: the case of *Myotragus balearicus* (Bate, 1909).' *Journal of Biogeography*. 30, 5, 771–781.

Braudel, F. (1988) *The Identity of France*. Vol. 1. London. English Trans. Siân Reynolds. Collins.

Braudel, F. (1992) *The Mediterranean and the Mediterranean world in the age of Philip II*. London. Collins Harper. Abr. Edn. Richard Ollard. Trans Siân Reynolds.

Broncano, M., Retana, J. and Anselm, R. (2005) 'Predicting the recovery of *pinus halepensis* and *quercus ilex* forests after a large wild fire in NE Spain.' *Vegetatio*. 180, 1, 47–56.

Brunet Estarellas, P. (1991) 'Aproximació a l'estudi de la propietat comunal a l'illa de Mallorca.' *Estudis Baleàrics*. 40, 99–112.

Brunhes, J. (1947) 'En Mallorca y Menorca.' *Estudios Geográficos*. 28, 545–560.

Bujosa, F. (1998) 'Amazing health rates in turn of century Majorca.' *Dynamis*. 18, 233–50.

Buswell, R. J. (1996) 'Tourism in the Balearic Islands.' In: Barke, M., Towner, J. and Newton, M. T. (eds) *Tourism in Spain: critical issues*. CAB International. Wallingford, 309–340.

Buswell, R. J. (2011) *Mallorca and tourism: history, economy and environment*. Bristol. Channel View Publications.

Butzer, K., Mateu, J., Butzer, E. and Kraus, P. (1985) 'Irrigation agrosystems in eastern Spain: Roman or Islamic origins?' *Annals of Association of American Geographers* 74 (4): 479–509.

Calatayud Llorca, P. (1991) *Un siglo de joyería y bisutería española, 1890–1990: exposición realizada por el Institut Balear de Diseny, Conselleria de Comerç i Indústria*. Palma. Conselleria de Cultura, Educació y Esports del Govern Balear.

Can Ontiveros, M. A. (2004) 'Ciutat romana de Palma; hipòtesis sobre el seu traçat urban i restes arqueològiques.' In: *Les ciutats romanes de llevant peninsular i les illes Balears*. Barcelona. Portic, 191–238.

Cañellas, N. (1993) *L'aigua, el vent, la sang: L'ús de les forces tradicionals a Mallorca*. Palma. Documenta Balear.

Cañellas Serrano, N. S. (1997) *El paisatge de l'Arxiduc*. Palma. Institut d'estudis Baleàrics i Consellaria Medi Ambient, Ordinació de territori i litoral.

Cantarellas Camps, C. (1994) *La Roqueta: una indústria ceràmica en Mallorca, 1897–1918*. Palma. Olañeta.

Carbonero Gamundi, M. A. (1983) 'Terrasses per al cultiu irrigat i distribució social de l'aigua a Banyalbufar.' *Documents d'anàlisi geogràfica*, 4, 31–68.

Carbonero Gamundi, M. A. (1984) 'L'origen i morphologia de les terrasses de cultiu a Mallorca.' *Boletin de la societe arqueològica Lul·liana*. 40, 91–100.

Carbonero, M. A. (1987) 'Estructura rural del sol rural.' In: Pie, R. (ed.) *Estudi sobre les parcel·lació, assentaments i nulcei des població el sol urbanitzable i no urbanitzable*. Palma. Ajuntament de Palma.

Carbonero Gamundi, M. A. (1992) *L'espai d'aigua. Petita hidràulica traditional a Mallorca*. Palma. Consell de Mallorca.

Carbonero Gamundi, M. A. (1991) 'Estructura rural i indústria a Palma, 1820–1930.' In: Manera, C. and Petrus Bey, J. M: *Del taller a la fàbrica*. Palma. Ajuntament de Palma, 91–100.

References

Casanova y Todolí, U. de and Lopez Bonet, J. E. (1986) *Diccionario de terminos historicos del reino de Mallorca (s. XIII–XVIII)*. Palma. Institut d'Estudis Baleàrics.

Casanovas, M-A. (2005) *L'economia balear (1898–1929)*. Palma. Edicions Documenta Balear.

Casey, J. (1999) *Early Modern Spain*. London. Routledge.

Cateura Bennasser, P. (1997) *Mallorca en el segle XIII*. Palma. El Tall.

Catlos, B. A. (2004) *The Victors and the Vanquished: Christians and Muslims of Catalonia and Aragon, 1050–1300* (Cambridge Studies in Medieval Life & Thought: Fourth Series.) Cambridge University Press. Cambridge.

Cau, M. A. and Chavez, M. E. (2003) 'El fenómeno urbano en Mallorca en época romana: los ejemplos de Pollentia y Palma.' *Mayurqa*, 29, 29–49.

Cela Conde C. J. (1979) *Capitalismo y campesinado el la isla de Mallorca*. Madrid. Siglo veintiuno de España editores.

Cerda, D. (2002) *Bocchoris. El mon clàssic a la badia de Pollença*. Col·lecció quaderns de patrimonia cultural, 8. Departement de Cultura. Palma. Consell de Mallorca.

Chadwick, A. M. (2004) '"Geographies of sentience" – an introduction to space, time and place.' In: Chadwick, A. M. (ed.) Stories from the landscape: archaeologies of inhabitation. Oxford. *BAR International Series* 1238, 1–31.

Cherry, J. F. (1990) 'The first colonisation of the Mediterranean Islands: a review of recent research.' *Journal of Mediterranean Archaeology* 3/2, 145-221.

Ciriacono, S. (2006) *Building on water: Venice, Holland and the construction of the European landscape in early modern times*. Oxford. Berghahn.

Cirer Costa, J. C. (2009) *La invenció del turisme de masses a Mallorca*. Palma. Documenta Balear.

Cohn, S. K. (2003) *The Black Death transformed: disease and culture in Early Renaissance Europe*. London. Hodder Arnold.

Col·lectius Pagès (2000) 'La rururbanització observada des d'un altre angle. Mallorca en venda al diari "Truque. "'. *Estudis Baleàrics*. 68–69, 171–186.

Colas, J-L. (1967) *The Balearics; islands of enchantment*. London. George Allen and Unwin.

Collins, R. (2004) *Visigothic Spain, 409–711*. Oxford. Blackwell.

Consell de Mallorca (1993) *Catàleg dels antics camins de la Serra de Tramuntana*. Palma. Consell de Mallorca.

Consell de Mallorca (2002) *Jaume II i les ordinacions de l'any 1300*. Catàleg de l'exposició. Palma. Departament de Cultura.

Consellaria de medi ambient, ordenació del territori i litoral (1997) *Directrices de ordenació territorial (DOT) – analysis y diagnostico*. Palma. Govern Balear.

Conselleria de comerç i indústria (1992) *Llibre blanc de la indústria a les Illes Balears. Primera part; anàlisi de la situació industrial*. Palma. Govern Balear.

Constable, O. R. (1994): *Trade and traders in Muslim Spain: the commercial realignment of the Iberian peninsula, 900–1500*. Cambridge. Cambridge University Press.

Constable, O. R. (2004) *Housing the Stranger in the Mediterranean World: Lodging, Trade, and Travel in Late Antiquity and the Middle Ages*. Cambridge. Cambridge University Press.

Cosgrove, D. (1990) 'An elemental division: water control and engineered landscape.' In: Cosgrove, D. and Petts, G. (1990) *Water, engineering and landscape*. London. Belhaven Press, 1–11.

Cosgrove, D. and Daniels, S. (1988) *The iconography of landscape; essays on the symbolic representation, design and use of past environments*. Cambridge. Cambridge University Press.

Cossens, N. (ed.) (2006) *England's landscapes*. 8 vols. London. English Heritage and Harper Collins.

Crawford Flitch, J. E. (1911) *Mediterranean moods*. London. Grant Richards.

Crespí M, Ferra Martorell, R. and Severa Sitjes, B. (1997) *Les Possessions de Porreres; estuí històric*. Porreres. Ajuntament de Porreres.

Bautista Dameto, J. and Mut, V. (1719) *The ancient and modern history of the Balearick Islands: or the Kingdom of Mallorca, which comprehends the islands of Majorca, Minorca, Yvica, Formentera and others*. London. Innys. Trans. from Spanish by C. Campbell.

Davis, M. H. L. A. (2002) 'Putting meat on the bone: an investigation into palaeodiet in the

Balearic Islands using carbon and nitrogen stable isotope analysis.' In: Waldren, W. H. and Ensenyat, J. A. *World Islands in Prehistory: Insular Investigations. Vth Deia International Conference of Prehistory*. BAR International Series 1095. Oxford, 198–216.

De Berard, G. (1789) *Viaje a las villas de Mallorca, 1789*. (new edn 1983) Palma. Ajuntament de Palma.

De Soto i Company, R. (1990) 'Repartiment and repartiments: l'ordinació d'un espai de colonició feudal a la Mallorca in de segle XIII.' In: *De Al Andalus a la sociedad feudal: los repartimentos bajo medieval*. Barcelona. CSIC, 1–51..

Deyá Bauzá, M. (1997) *La manufactura de la llana a la Mallorca del segle XV*. Palma. El Tall.

Deyá Bauzá, M. J. (1998) *La manufactura de la llana a la Mallorca moderna (Segles XVI–XVII)*. Palma. El Tall.

Deyá Bauzá, M. J. (2000) 'Un exemple de la indústria rural tèxtil al Pla de Mallorca: Algaida durant l'Antic Régim.' In: *Algaida, cinques jornades d'estudis locals*. Llucmajor. Mancomunitat Pla de Mallorca, 73–78.

Deyá Bauzá, M. J. (2002) 'Repercussions ecoambientals dels models econòmics a Mallorca: les seves relacions amb l'estructura social (segles XIII–XVII).' *Estudis d'història econòmica*. 19, 67–86.

Dodds, R. (2007) 'Sustainable tourism and policy implementation: lessons from the case of Calvià, Spain.' *Current issues in tourism*. 10, 4, 296–322.

Doenges, N. (2005) *Pollentia: a Roman colony on the island of Mallorca*. BAR report S1404. Oxford. Archaeopress.

Doenges, N. (2007) 'Pollentia: a Roman city on the island of Mallorca, Spain.' William L. Bryant Foundation. Dartmouth College excavations. www.dartmouth.edu/~classics/oldsite/pollentia.

Dorale, J. *et al.* (2010) 'Sea-Level Highstand 81,000 Years Ago in Mallorca.' *Science*. 12 February 2010: 860–863.

Dorrian, M. and Rose, G. (2003) *Landscape and politics*. London. Black Dog Publishing.

Duran, D. (1980) *Porto Cristo: Societat i cultura*. Manacor. n. p.

Duran, E. (1979) 'La crisi rural mallorquina dels segles XV i XVI.' *Estudis d'història agrària*. 3, 53–77.

Duran Jaume, D. (2003) *Producció i crisi arrosa de Sa Pobla i Muro*. Pollença. El Gall. (2nd Edn).

Duran Jaume, D (2004) *El mode de vida del pescador de Cala de Manacor*. Pollença. El Gall (2nd Edn).

Dyson, S. L. (2003) *The Roman countryside*. London. Duckworth.

Elliott, J. H. (1990) *Imperial Spain, 1468–1716*. (2nd Edn). Harmondsworth. Penguin.

Elwell, C. J. L. (1995) *The greater aphrodisiad or Majorca*. London. Ferrington.

Emery, F. (1969) *The World's Landscapes – Wales*. London. Longmans.

Escartín, J. M. (1991) *El procés d'industrialització a Esporles, 1830–1960*. Esporles. Ajuntament d'Esporles.

Escartín, J. M. *et al.* (1996) *El sector de la madera y el mueble en Baleares*. Palma. Gabinete Tecnico CC. OO Illes Balears.

Escartín, J. M. (2001) *La ciutat amuntegada: Indústria del calçat, desenvolupament urbà i condicions de vida en la Palma contemporània (1840–1940)*. Palma. Documenta Balear.

Eugénio, M. and Lloret, F. (2006) 'Effects of repeated burning in Mediterranean communities of the NE Iberian peninsula.' *Journal of Vegetation Science*. 17, 6, 755.

Fernandez-Armesto, F. (1987) *Before Columbus: exploration and colonisation from the Mediterranean to the Atlantic, 1229–1492*. London. Macmillan Education.

Ferragut Bonet, J. (1974) 'La demortizacion de Mendizábal en Mallorca.' *BCOCIN*. 5, 125–179.

Ferrer Febrer, A. and Carvajal Mesquida, A. (2003) *Evolució urbana de Manacor (1600–1944)*. Manacor. Ajuntament de Manacor.

Ferrer Flórez, M. (1974) *Población y propiedad en la cordillera septentrional de Mallorca*. Vols 1 and 2. Palma. Instituto de Estudios Baleáricos.

Ferrer Flórez, M. (1981) *Agricultura en el siglo XVIII: aspectos agrarios de la cordillera septentrional de Mallorca*. Palma. Escuela Universitaria de Estudios Empresariales.

References

Ferrer Flórez, M. (1983) *El trabajo agrario en los siglos XVI y XVII: formas de explotación agraria en la cordillera septentrional de Mallorca*. Palma. Escuela Universitaria de Estudios Empresariales.

Florit Alomar, F. (1983) *Les transformacions del paisatge rural i de la propietat de la terra a la comarca de Pla de Mallorca. Ss. XIX–XX*. Departamento de Geografia. Palma. Universidad de Palma de Mallorca.

Font Jaume, A. *et al.* (1995) *Una villa romana al Pla de Mallorca*. Palma. Lleonard Muntaner.

Font Obrador, B. (2000) *Jaume II de Mallorca i les ordinacions de les pobles: gènesis de la vila de Llucamajor*, Llucmajor. Ajuntament de Llucmajor.

Fornos, J. J. (1995) 'Enquadrament geològic, evolució estructural i sedimentologia de S'Albufera de Mallorca.' In: Martínez Taberna, A. and Mayol Serra, J. (eds) *S'Albufera de Mallorca*. Palma. Moll.

Foxhall, L. (2003) 'Cultures, landscapes and identities in the Mediterranean World.' *Mediterranean Historical Review*. 18, 2, 75–92.

Gabriel Fiol, M. (2002) *Mancor, Massenella, Biniarroi, Biniatzent; noticies històrics 1601–1800*. Mancor. Ajuntament de Mancor de la Vall.

Gallois, L. (1908) *Régions naturelle et noms de pays*. Paris. Colin. (reprint, CTHS, 2008).

Garau Febrer, T. (1995) *Historia de Son Macià*. Palma. Lleonard Muntaner.

Garcia Riaza, E. (2003) 'Las ciudades romanas de Mallorca y su diversidad.' *Mayurqa*. 29, 73–83.

García-Delgado Segués, C. and Garau Alemany, L. (2000) *Las raíces de Palma: los mil primeros años de la construcción de una ciudad: de la colonia romana a la medina musulmana*. Palma. Olañeta.

García-Delgado Segués, C. (1979) 'Ciutat de Mallorca – evolution y permanencia del centro historico.' *La construccion de la ciudad*. 13, 6–24.

García-Delgado Segués, C. (1998): *La Casa Mallorquina. Influencias de Roma, el Islam y catalunya*. Palma. Olañeta.

Gil, L., Manuel, C. and Diaz-Fernandez, P. (2003) *La transformación histórica del paisaje forestal en las Islas Baleares*. Madrid. Instituto Forestal National.

Gilbert, E. W. (1934) 'The human geography of Mallorca.' *Scottish Geographical Magazine*. 3, 31–147.

Ginard, D. (1999) *L'economia Balear (1929–1959)*. Palma. Edicions Documenta Balear.

Ginard Bujosa, A. (1995) *Evolució històrica de l'abastament d'aigua a Palma (1800–1995): un debat permanent*. Palma. Emaya, Ajuntament de Palma.

Gines, J. and Gines, A. 1989) 'El karst de las isles Baleares.' In: *El karst en Espana*. Madrid. Sociedad Española de Geomorfologia. Monograph 4, 161–174.

Glick, T. (2000) 'Reading the *repartimentos*; modeling settlement in the wake of conquest.' In: Meyerson, M. D. and English, E. D. (eds) *Christians, Muslims and Jews and Early Modern Spain*. Notre Dame, Indiana. University of Notre Dame Press, 20–39.

Glick, T. and Kirchner, H. (2000) 'Hydraulic systems and technologies of Islamic Spain: History and Archaeology.' In: P. Squatriti (ed.) *Working with Water in Medieval Europe: Technology and Resource-Use*. Leiden, 267–329.

Glick, T. F. (1995) *From Muslim fortress to Christian castle: social and cultural change in medieval Spain*. Manchester. Manchester University Press.

Glick, T. F. (2002) 'Tribal landscapes of Islamic Spain – history and archaeology.' In: Howe, J. and Wolfe, M. (eds) *Inventing medieval landscapes: sense of place in medieval Europe*. Gainsville. University of Florida Press, 113–135.

Glick, T. F. (2007) *Paisajes de conquista; cambio cultural y geográfico en la España medieval*. València. Universitat de València. Spanish Trans. from Glick, 1995, J. Torró.

Goldring, D. (1946) *Journeys in the sun. Memories of happy days in France, Italy and the Balearic Islands*. London. Macdonald.

Gomez Mendoza, J. and Mata Olmo, R. (2006) 'Paisajes forestales españoles y sostenibilidad: tópicos y realites.' *Areas*, 25, 13–30.

Gomez Ortiz, A. (1992) 'Les muntanyes de les illes Balears.' In: *Geogràfic general del paises Catalans vol 1. El clima i el relleu*, 282–286.

González de Molina, M. (2001) 'The limits of agricultural growth in the nineteenth century: A case study from the Mediterranean world.' *Environment and History*. 7, 4, 479–499.

González Pérez, M. (2003) 'La pérdida espacios de identidad y la construcción de lugares en el paisaje turístico de Mallorca.' *Boletín de la Asociación Geografos Españoles* 35, 137–152.

Gorrias i Duran, A. and Terradas Jofre, R. (2005) 'L'aprofitament pre-industrial de l'aigua de la Granja. La Font Major de la Granja i la Font d'en Baster dues fontes, La Mateixa aigua.' In: *IV Congres International de molinologia. Actes.* vol 1, 313–367. Palma. Consell de Mallorca.

Govern Balear (1987) *Espacios naturales de Baleares; evaluacion de 73 areas para su proteccion.* Palma. Conselaria d'obres públiques i ordinació del territori.

Graves, R. and Hogarth, P. (1965) *Majorca observed.* London. Cassell.

Grimalt Gelabert, M. (1989) 'Les inundacions històriques de Sa Riera.' *Treballs de Geografia.* 42, 21.

Grimalt Gelabert, M. and Blazquez, M. (1989) 'El mapa de marjades de la Sierra de Tramuntana de Mallorca.' *Treballs de Geografia.* 42, 43–7.

Grimalt Gelabert, M. and Rodriguez Gomila, R. (2001) 'Toponims i geografia de les possessions del litoral del terme de Manacor.' *Jornades d'estudis locals de Manacor* 1, 59–73.

Grimalt Gelabert, M. *et al.* (1991) *Libro-guia de las excursines de las VII Jornadas de Campo de Geografica Fisica.* Palma. Universitat de les Illes Balears.

Grimalt Gelabert, M. *et al.* (2003) *Paisatge i pedra en sec a Mallorca. La serra de Tramuntana de Mallorca, les muntanyes construïdes.* Palma. Consell de Mallorca-FODESMA, Programa de catalogació i anàlisi dels espais de marjades de Mallorca.

Grimalt Gelabert, M., Laita Ruíz de Asua, M. and Ruíz Perez, M. (2001) 'Pautes espacials i temporals de distribució d'aiguades intenses al Llevant de Mallorca (1930–1995).' *Jornades d'estudis locals de Manacor,* 1, 29–39.

Grosske Fiol, E. (1978–9) 'Diez años de desamortizacion en Mallorca (1855–65).' *Trabajos de Geográfia.* 35, 93–103.

Grove, A. T. and Rackham, O. (2001) *The nature of Mediterranean Europe – an ecological history.* New Haven and London. Yale University Press.

Guererro Ayuso, V. M. (1998) *La Mallorca prehistòrica. Des dels inicis al bronze final.* Palma. El Tall.

Guerrero Ayuso, V. M. (2001) 'The Balearic Islands: prehistoric colonisation of the furthest islands from the mainland.' *Journal of Mediterranean Archaeology* 14, 2, 136–158.

Guichard, P. (1998) 'L'aprofitament de l'aigua.' In:. *Enciclopèdia Catalana: historia, política, societat i cultura dels països catalans.* Barcelona.

Gutirrez Lloret, S. (1987) 'Elementos de urbanismo de la capital de Mallorca: funcionalidad espacial.' In: Rossello Bordoy, G. (ed.) *Les Illes orientals d'al-andalus. V Jornades d'estudis històrics locals.* Palma. Institut d'estudis Baleàrics. . 205–224.

Hardin, G. (1968) 'The Tragedy of the Commons.' *Science,* 162, 1243–1248.

Harris, W. V. (2005) *Rethinking the Mediterranean.* Oxford. Oxford University Press.

Harrison, J. and Corkhill, D. (2004) *Spain: a modern European economy.* Aldershot. Ashgate.

Hernandez Gasch, J. and Aramburu-Zabala, J. (2005) 'Origen i evolució dels poblats talaiòtics.' In: Sals Burguera, M. (Coord): *Mirant el passat; curs de prehistòria de Mallorca.* Papers de Sa Torre 63. Patronat De L'escola municipal de Mallorca, Manacor.

Hewitt, R. (2010) *Map of a nation: a biography of the Ordnance Survey.* London. Granta.

Hillgarth, J. N. (1976) *The Spanish Kingdoms 1250–1516.* 2 vols. Oxford. Clarendon Press.

Hillgarth, J. N. (1991) *Readers and books in Majorca, 1229–1550.* Vol. 1. Paris. Editions du Centre National de la Recherché Scientifique.

Hillgarth, J. N. (2003) *Spain and the Mediterranean in the later Middle Ages: studies in political and intellectual history .* Aldershot. Ashgate. (Vorarium Edn).

Horden, P. and Purcell, N. (2000) *The Corrupting Sea: a study in Mediterranean history.* Oxford. Blackwell.

Horden, P. and Purcell, N. (2006) 'The Mediterranean and "the New Thalassology".' *The American Historical Review,* 111, 3, 722–740.

Hoskins W G (1955): *The making of the English landscape.* London. Hodder and Stoughton.

Hoskins, W. G. (1988) *The making of the English landscape.* London. Hodder and Stoughton. With introduction and commentary by Christopher Taylor.

Hunt, J. D. (2003) 'Taking place: some preoccupations and politics of landscape study.' In: Dorrian, M. and Rose, G. (eds) *Landscape and politics.* London. Black Dog Publishing,

122–129.

Idrisi, Z. (2005) 'The Muslim agricultural revolution and its influence on Europe.' Foundation for Science, Technology and Civilisation: www.muslimheritage.com/uploads.AgriRev2pdf,5.

Jackson, J. B. (1984) *Discovering the vernacular landscape*. New Haven. Yale University Press.

Jover Avellà, G. (2001) 'Ingresos y estrategias patrimoniales de la nobleza durante la crisis de seiscientos, Mallorca, 1600–1700.' In: *Congreso de historia economica, Zaragosa, Sept. 2001. Papers in session 15*, 105–27.

Jover Avellà, G. and de Soto, R. (2002) Colonización feudal y organización del territorio, Mallorca, 1230–1350.' *Revista de Historia Económica*. 20 (3): 437–75.

Jover Avellà, G. and Morey, A. (2003) 'Les possessions mallorquines: una modalitat d'organització de l'espai agrari i de l'explotació del treball.' In: Congost, R., Jover, G. and Biagioli, G. (eds) *L'organització de l'espai rural. Masos, possessions, poderi*. Girona. CCG Edicions, Col·lecció Estudis, 5, 127–238.

Juan, J. (1981) 'La production de aceite en Mallorca durante la Edad Moderna y su papel en la económia mallorquina.' *Bolletí de la Societat Arqueologica Lul. liana*, 832–33.

Juan, J. (1990) 'Crisis de subsistència i aprovisionament blader de Mallorca durant el segle XVIII.' *Randa*, 26

Kamen, H. (1991) *Spain, 1469–1714: a society of conflict*. 2nd Edn. London. Macmillan.

Keay, S. J. (1988) *Roman Spain*. London. British Museum Publications.

Kennedy, H. (1996) *Muslim Spain and Portugal*. London. Longman.

Kent, M., Newnham, R. and Essex, S. (2002) 'Tourism and sustainable water supply in Mallorca: a geographical analysis.' *Applied Geography*. 22 (4): 351–374.

King, R. (1997) 'Introduction: an essay on Mediterraneanism.' In: King, R., Proudfoot, L. and Smith, B. (eds) *The Mediterranean: environment and society*. London. Arnold. 1–11.

King, R., Proudfoot, L. and Smith, B. (eds) *The Mediterranean: environment and society*. London. Arnold.

Kirchner, H. (1994) 'Espais irrigats andalusins a la serra de Tramuntana.' *Afers* 18, 313–335.

Kirchner, H. (1997) *La construcció de l'espai pagès a Mayurqa: les valls de Bunyola, Orient, Coanegra i Alaró*. Palma. U. I. B. (Published form of PhD thesis).

Kirchner, H. (2005) 'Molins hidràulics andalusí a Mallorca i Eivissa.' *IV Congres Internacional de molinolgia . Actes* Vol 1, 239-63.

Kirchner, H. and Retamero, F. (2004) 'Cap a una arqueologia de la colonització. La subversió feudal de l'espai rural a les Illes.' *Avenc*. 290, 40–5.

Kostof, S. (1999) *The City Shaped*. London. Thames and Hudson.

Ladaria Benares, M. D. (1992) *El ensanche de Palma; planteadito del tema, problemática construcción y valoración de un nuevo espacio urbano, 1868–1927*. Palma. Ajuntament de Palma.

Lambert, A. (1971) *Making of the Dutch landscape: historical geography of the Netherlands*. London. Academic Press.

Landes, D. S. (1998) *The wealth and poverty of nations*. New York. Norton.

Lenček, L. and Bosker, G. (1999) *The beach: a history of paradise on earth*. London. Pimlico.

Lewis, B. (1960) *The Arabs in History*. London. Hutchinson University Library.

Llabres Bernat, A. (1997) *El paisatge a les Balears*. Conselleria de Medi Ambient, Ordinació del Territori i Literal. Palma. Govern Balear.

Lluch i Dubon, F. (1997) *Geografia de les illes Balears*. Palma. Lleonard Muntaner.

Lopez Bravo, P. (2003) Impacts of tourism on Majorca: the new colonisation. Unpublished MA thesis. Bournemouth University. www. du. se Consulted 8/9/2009.

Lord, P. (ed.) (1979) *A Moorish Calendar – from the Book of Agriculture of Ibn Al-Awam*. Wantage. Black Swan Press.

Lourie, E. (1976) 'Free Moslems in the Balearics under Christian rule in the 13th Century.' *Speculum*. 45, 624–649.

Lucena, M., Fontenia, J. M. *et al*. (1997) *Palma, Guia de arquitectura*. Palma. Col. legi Oficial d'arquitectes de Balears.

Luis Gómez, A. (1980) 'El geógrafo español. Aprendiz de brujo? Algunos problemas de la geografía del paisaje.' *Geo Crítica*. 5, 25.

References

Luis Gómez, A. (1984) 'Geografia social y geografia del paisaje.' *Geo Crítica*. 10, 49.

Manera, C. (1983) 'Notas sobre exportaciones mallorquinas de aciete a Marsalla en la secunda mitad del Setecientos.' *Jornades d'Estudis Històrics Locals IV*. Palma.

Manera, C. (1988) 'Manufactura textil y comercio en Mallorca, 1700–1830.' *Revista de Historia Económica*. 6, 3, 523–555.

Manera, C. (1991) 'Mercado, producción agrícola y cambio económico en Mallorca durante el siglo XVIII.' *Revista de Historia Económica*, 9, 1, 69–101.

Manera, C. (2001) *Història del creixement econòmic a Mallorca, 1700–2000*. Palma. Lleonard Muntaner.

Manera, C. (2003) *El model històric de creixement a les Illes Balears: un intent de teorització; discurs de recepció del Premi Catalunya d'economia 2003*. Palma. Lleonard Muntaner.

Manera, C. (2006a) 'La força industrial a les Illes Balears; anàlisi del passat i projecció cap al futur.' *VII Jornades d'Estudis Locals*. Inca, pp. 11–35.

Manera, C. (2006b) *La riquesa de Mallorca: una historia econòmica*. Palma. Lleonard Muntaner.

Manera, C. and Escartín, J. M. (1992) *El sector metalúrgico en la Mallorca contemporanea, 1700–1950*. Palma. CC. OO. Illes Balears. Gabinete Tècnico.

Manera, C. and Petrus Bey, M. A. (1991) 'El sector indústrial en el creixement econòmic de Mallorca, 1780–1985.' In: Manera, C. and Petrus Bey (eds) *Del taller a la Fàbrica*. Palma. Ajuntament de Palma. 13–58.

Manera, C. and Roca Avellà, J. (1995) 'Patrimoni industrial perdut, cultura industrial esvaïda. Notes sobre el cas de Mallorca.' In: G. Rossello Bordoy (ed.) *Actes de III congres: el nostre patrimoni cultural; el patrimoni tudat (1836–1994). Bolletí de la Societat Arqueològica Lul·liana*. Palma, 185–208.

Marin, M. (ed.) (1998) *The formation of al-Andalus. Part 1: History and Society*. Aldershot. Ashgate.

Markham, C. R. (1908) *The story of Majorca and Minorca*. London. Smith, Elder & Co.

Mas, A. and de Soto i Company, R. (2004) 'Un regne dins la mar. El procés migratori català i l'extinció da la població indígena a Mallorca.' *Avenc* 290, 35–9.

Mas Forners, A., Rosselló Bordoy, G. and Rosselló Vaquer, R. (1999) *Historia d' Alcúdia de l'epoca Islamica a la Germania*. Alcúdia. Ajuntament d'Alcúdia.

Mateos Ruiz, R. M. (2010) 'Deslizamientos históricos en la isla de Mallorca: Biniarroi, 1721 y Fornalutx, 1924.' In: Mayol, J. *et al.* (eds) *Homenatge a Bartomeu Barceló i Pons, geògraf*. Palma. Lleonard Muntaner, 215–231.

Matless, D. (1993) 'W. G. Hoskins and the English culture of landscape.' *Rural History*. 4 (2): 187-208.

Mayol, J. *et al.* (eds) (2010) *Homenatge a Bartomeu Barceló i Pons, geògraf*. Palma. Lleonard Muntaner, 215–31.

McNeill, J. R. (1992) *The mountains of the Mediterranean world*. Cambridge. Cambridge University Press.

Meinig, D. W. (1979) *The interpretation of ordinary landscapes*. Oxford. Oxford University Press.

Merino Santisteban, J. and Torres Orell, F. (2000) 'El poblament rural de Manacor de la conquesta romana la islamització.' *Jornades d'estudis locals de Manacor*. 1, 143–57.

Metvejevic, P. (1999) *Mediterranean: a cultural landscape*. Berkeley. University of California Press.

Mico Perez, R. (2005) 'Cronología absoluta y periodización de la prehistoria de las Islas Baleares.' *British Archaeological Reports International Series*. 1373. Oxford.

Miranda González, M. A. (2001) 'Inmigration y cohesion social en Calvià, Mallorca.' *Revista Electrónica de Geografía y Ciencias Sociales*. 94, 21.

Moll i Blanes, I. (1996) 'El paisatge agrari de Marratxí a 1818.' *Jornades d'estudis locals a Marratxí*. 1, 53–73.

Moll i Blanes, I. (1997) 'El paisatge de Marratxí: un revisió.' *2 Jornades d'estudis locals a Marratxí*. Marratxí. Ajuntamant de Marratxí, 57–64.

Moll i Blanes, I. (1997–8) 'El espació rural del municipio de Palma en 1816.' *Mayurqa*. 24, 79–88.

Moll i Blanes, I., Segura, A. and Suau, J. (1983) *Cronologia de les crisis demogràfiques a Mallorca, segles XVIII–XIX*. Palma. Institut d'Estudis Baleàrics.

Montaner, P. de and Le-Senna, A. (1981) 'Explotació d'una possessió mallorquina durant la primera meitat del segle XVI: Son Sureda (Marratxí).' *Receques*. 11, 107–24.

Montufo, A. M. (1998) 'The use of satellite imagery and digital image processing in landscape archaeology. A case study from the island of Mallorca, Spain.' *Geoarchaeology*. 12, 1, 71–85.

Moranta Jaume, L. (2007) 'Hipotesis de la existencia de un theatre romano en Palma de Mallorca.' www.palma.infotelecom.es/~moranta.

Morey Tous, A. (2001) 'Aproximació a la situació econòmica de Manacor a partir d'integració de 1800.' In: *Manacor: cultura i territori: I Jornades d'estudi locals de Manacor: 5 i 6 de maig*. Manacor. Ajuntament de Manacor.

Morley, J. and Plant, K. P. (1963) *Minor railways and tramways in Eastern Spain*. Birmingham. Birmingham Locomotive Club.

Morro Veny, G. (2003) *Capdepera medieval; segles XIII and XIV*. Palma. Edicions Documenta Balear.

Muir, R. (1981) *Reading the landscape*. London. Michael Joseph.

Mulet i Ramis, B., Rosselló i Vaquer, R. and Salom i Sancho, J. (1995): *La crisi de la vila de Sineu (segle XV)*. Sineu. Ajuntament de Sineu.

Ne'eman, G. and Louis Trabaud, L. (2000) *Ecology, biogeography and management of* Pinus halepensis and P. Brutia *forest ecosystems in the Mediterranean basin*. Leiden. Backhuys Publishers.

Nogué, J. (2005) 'Nacionalismo, territorio y paisaje en Cataluña.' In: Ortega Cantero, N. (ed.) *Paisaje, memoria histórica e identidad nacional* Collección de Estudios, 102. Fundacion Duques de Soria. Madrid. UAM Edicones, 147–70.

Oliver Costa, C. (2003) 'La parròquia del Carme de Porto Cristo.' *Jornades d'estudis locals de Manacor*. 2, 301–3.

Oliver Servera, L. (2005) 'Les datacions radiocarboniques als Closos de Can Gaià (Mallorca).' *Mayurqa*. 30, 1, 247–62.

Ordinas Marce, G. (1990) 'L'exploitació de l'Albufera als segles XVII, XVIII, XIX a traves dels arrendaments.' In: *1 Jornades d'estudis locals d'Alcúdia*. Alcúdia. Ajuntament d'Alcúdia. 51–62.

Ordinas Marcé, G. (1995) *Els forns de calç a Santa Maria del Camí*. Ajuntament de Santa Maria del Camí.

Orfila, M. (ed.) (2000) *El forum de Pollentia. Memòria d'excavacions realitzades entre els anys 1996 i 1999*. Alcúdia. Ajuntament de Alcúdia.

Orfila Pons, M. (1993) 'Construcciones rurales romanas en Mallorca.' In: Padro, J. *et al.* (eds) 'Homenatge a Miguel Tarradell.' *Estudis Universitaris Catalans XXIX*, 793–805.

Orfila Pons, M. (2005) 'Romanisation of the Illes Balears through archaeology.' In: Tugores Truyol, F. (Co-ordinator) *El mon romà a les illes Baleares*. Barcelona. Catalogue of exhibition. Fondació 'la Caixa'.

Orfila Pons, M., Arribas Palau, A. and Doenges, N. (1999) 'El fòrum de la ciutat romana de Pollentia: estat actual de les investigacions.' *I Jornades d'Estudis Locals d'Alcúdia: 13 i 14 de novembre de 1998*. Alcúdia. Ajuntament d'Alcúdia.

Ortega Cantero, N. (ed.) (2004) *Naturaleza y cultura del paisaje*. Collección de Estudios, 91. Fundación Duques de Soria. Madrid. UAM Edicones

Ortega Cantero, N. (ed.) (2005) *Paisaje, memoria histórica e identidad nacional* Collección de Estudios, 102. Fundacion Duques de Soria. Madrid. UAM Edicones

Ortega Cantero, N. (ed.) (2010) 'El lugar del paisaje en la geografía moderna.' *Estudios Geográficos*. 71 (269).

Palanques, M. L. and Calvo, M. (2011) Cartography, cadastre and surveying and the development of cities during nineteenth century Spain. www.fig.net/pub/fig2011/papers/ts09g/ts09g_palanques_calvo_5029.pdf. Consulted 4/8/2011.

Papatheodorou, A. (2004) 'Exploring the evolution of tourism resorts.' *Annals of Tourism Research*. 31 (1): 219–237.

Pastor Sureda, B. (1976) Maria de la Salut: un exemple de camí de paisatge dins el Pla de Mallorca. *Mayurqa*. 16.

Pastor Sureda, B. (1977–8) 'Les colònies agrícoles del segle XIX a Mallorca.' *Mayurqa*. 17, 175–77.

Pastor Sureda, B. (2000) 'La fil. loxera al Pla de Mallorca.' In: *Algaida, cinques jornades d'estudis locals*. Llucmajor. Mancomunitat Pla de Mallorca, 147–49.

Patton, M. (1996) *Islands in time; island sociogeography and Mediterranean prehistory*. London. Routledge.

Penya Barceló, A. (1991) 'La ciutat i les manufactures; aspectes de la indústria urbana al segle XIX.' In: Manera, C. and Petrus Bey, M: *Del Taller a la Fàbrica*. Palma. Ajuntament de Palma. pp 59–68.

Pericot García, L. (1972) *The Balearic Islands* (Ancient people and places series). London. Thames and Hudson.

Peterson A. F. (1979) War, politics and the Kingdom of Mallorca, 1621–1641. Ph. D thesis, Johns Hopkins University, Baltimore, Maryland.

Picazo Muntaner, A. (2001–2) 'Hisenda i parcel. lacions rustiques. La creació del nucleó de Son Servera.' *Estudis Baleàrics*. 70–71, 123–128.

Picornell, C. and Ginard, A. (1995) 'John Frederick Bateman.' In: Martínez, A. and Mayol, J. (eds) *L'Albufera de Mallorca*. Monografies de la Societat d'Història Natural de les Balears, 4. Palma. Moll. pp 39–46.

Picornell, C. and Picornell, C. (2008) *Apunts del Pla de Mallorca*. Pollenca. El Gall and Institut d'Estudis Balears.

Pieras Salom, G. (1995) *Inca 1872: carrers, finques, habitants i oficis*. Inca. Ajuntament d'Inca.

Pique, R. and Noguera, M. (2002) 'Landscape and management of forest resources in the Balearic Islands during II–I Millennium BCE.' In: Waldren, W. H. and Ensenyat, J. A. (eds) (2002) *World Islands in Prehistory: Insular Investigations. Vth Deia International Conference of Prehistory*. BAR International Series. 1095. Oxford, 292–300.

Poveda Sanchez, A. (1980) 'Introducción al estudio de la toponimia árabe-musulmana de Mayurqa según la documentación de los Archivos de la Ciutat de Mallorca, 1232–1278.' *Awraq*. 3, 75–102.

Poveda Sanchez, A. (1992) 'Mayurqa, un espació agrario de alquerias y rafales.' *Estudis d'historia economica*. 1, 5–11.

Presley, J. W. (2002) '"Frizzling in the sun": Robert Graves and the development of mass tourism in the Balearic Islands.' In: Robinson, M. and Anderson, H. C. (eds) *Literature and tourism*. London. Continuum.

Pryor, F. (2010) *The making of the British Landscape*. London. Allen Lane.

Pryor, J. (1988) *Geography, Technology and War*. Cambridge. Cambridge University Press.

Pujalte i Vilanova, F. (2000) 'El rumor sumergido de la isla de la calma.' In: Tugores Truyol, F. (Co-ordinator): *Welcome! Un siglo de turismo en las Islas Baleares*. Catalogue of exhibition. Fondació 'la Caixa'. Barcelona.

Pujalte i Vilanova, F. (2002) *Transports i comunicacions a les Balears durant el segle XX*. Palma. Documenta Balear. Quaderns d'història contemporània de les Balears.

Pujalte i Vilanova, F. (2006) 'Electricitat i el procés d'industrialització a Mallorca: algunes reflexions.' In: *Un segle de llum a Inca, 1905–2005*. Palma. Documenta Balear.

Quadrado, J. M. (1889–90) 'La ciudad de Mallorca en el siglo XV.' *Bolletin de la Societad Aqueologica Luliana*, 3.

Quintana Peñuela, A. (1979) *El systems urbano de Mallorca*. Palma. Moll.

Rackham, O. (1976) *Trees and woodland in the British landscape*. London. Dent.

Rackham, O. (2000) 'The medieval countryside of England: botany and archaeology.' In: Howe, J. and Wolfe, M. (eds) *Inventing medieval landscapes: senses of place in medieval Europe*. Gainsville. University of Florida Press, 13–32.

Rackham, O. and Moody, J. (1996) *The making of the Cretan landscape*. Manchester. Manchester University Press.

Ramis, D. and Alcover, J. A. (2001) 'Revisiting the earliest human presence in Mallorca, Western Mediterranean.' *Proceedings of the Prehistoric Society*. 67, 261–69.

Ramis, D. Alcover, J. A., Coll, J. and Trias, M. (2002) 'The chronology of the first settlement of the Balearic Islands.' *Journal of Mediterranean Archaeology*. 15, 1, 3–24.

Riera Frau, M. (1993) *Evolució urbana i topographia de Medina Mayurqa*. Palma. Ajuntament de Palma.

References

Riera Frau, M. (1994) Aigua i disseny urbà: Medina Mayurqa. *Afers* 18, 305–12.

Riera Frau, M. and Soberals Sagreras, N. (1992) *El sistema hidràulic d'Alaró (Mallorca)*. Ajuntament de la vila d'Alaro (reprinted from *Bolletin de la Societad Aqueologica Lul. liana*. 47, 61–73).

Richardson, J. S. (1996) *The Romans in Spain*. Oxford. Blackwell.

Ringrose, D. (1996) 'Spain, Europe and the "Spanish Miracle", 1700–1900.' Cambridge. Cambridge University Press.

Robledo, M. A. and Batle, J. (2002) Re-planning for tourism in a mature destination: a note on Mallorca.' In: Voase, R. (ed.) *Tourism in Western Europe: a collection of case histories*. Wallingford. CABI.

Roca Avellà, J. (1992) 'Modernització agrícola i desenvolupament industrial. El cas de Mallorca (1850–1950).' *Estudis Baleàrics* 43, 109–119.

Roca Avellà, J. (2006) *La Indústria a Mallorca* (segles XIX–XX). Palma. Documenta Balear. Quaderns d'història contemporània de les Balears.

Roca Avellà, J. (2006) *Llana, vapors, coto i negoci; aproximació a la indústria tèxtil de la Mallorca del vuits cents. El cas de Can Ribas (1850–1885)*. Palma. Edicions Documenta Balear.

Roca Avellà, J. (2010) *La Indústria mallorquina durant el franquisme (1939–75)*. Palma. Documenta Balear. Quaderns d'història contemporània de les Balears.

Roca Avellà, J. and Umbert, J. (1990) 'Economía y desarrollo industrial en Mallorca (1914–30): 'Apuntes de investigácion.' *Estudis d'historia economica*. 1, 93–112.

Rodríguez Alcalde, A. (1996) 'Aproximacion estadistica al paisaje humano en la prehistoria de Mallorca.' *Complutum* 6, 167–192.

Roman Quetglas, J. (2001) 'Aportació a l'estudí històric i artístic del convent de Sant Vicenç Ferrer. Segles XVI–XVII.' *Jornades d'estudis locals de Manacor*. 1, 385–401.

Roman Quetgles, J. (2005) 'Els jardins de Raixa.' *Bolletin de la Societad Aqueologica Lul. liana*, 61, 197–212.

Rosselló Bordoy, G. (1983) 'La evolución urbana de Palma en la antigüedad: 1. Palma roman.' *BCOCIN* 631, 121–39.

Rosselló Vaquer, R. (1977) *Historia de Campos, Vol 1 de la prehistòria al sigle XVI*. Campos. Ajuntament de Campos.

Rosselló Vaquer, R. (1996) 'Bunyola en el segle XV.' Bunyola. Col. lectiu Cultural Sitja. Bunyola.

Rosselló Vaquer, R. and Vaquer Bennàsar, O. (1991) *Historia de Manacor: El segle XVI*. Palma. Prensa Universitaria.

Rosselló Verger, V. (1964) *Mallorca: el sur y sureste*. Palma. Camara Oficial de Comercio, Industria y Navigacion de Palma de Mallorca.

Rosselló Verger, V. M. (1974) 'La persistència del castrato romano en el Mighorn de Mallorca.' *Estudios sobre centuriaciones romanas en España*. Universidad Autonoma de Madrid, 136-55

Rosselló Verger, V. M. (1981) 'Canvis de propritat i parcel. lacions al camp mallorquí entre els segles XIX i XX.' *Randa*. 12, 19–60.

Rosselló Verger, V. M. (2000) 'Felanitx, el medi i la gent.' *Jornades d'estudis locals de Felanitx*, 9–21.

Rowley, T. (2006) *The English Landscape in the Twentieth Century*. London. Continuum.

Ruggles, D. (2002) *Gardens, landscape and vision in the palaces of Islamic Spain*. New York. Penn State University Press.

Ruiz-Viñals, C. (2000) *L'urbanisme de la ciutat de Palma de Mallorca*. Palma. El Far de les Crestes.

Rullan Salamanca, O. (1998) 'De la cove de Canet al tercer "boom" turístic. Una primera aproximació a la geografia històrica de Mallorca.' In: *El Mediambient a les Illes Balears. Qui es qui?* Palma. Caixa de Balears 'Sa Nostra', 171–213.

Rullan Salamanca, O. (2002) *La construcció territorial de Mallorca*. Palma. Moll.

Rusiñol, S. (1889) *La Isla de la Calma*. Palma. Editorial Baleares. Eng Trans.

Sala, P. (2000) 'Modern forestry and enclosure: elitist state science against communal management and unrestricted privatisation in Spain, 1855–1900.' *Environment and History*, 6, (2):151–68.

Salas Colom, A. (1992) *El turismo en Mallorca: 50 años de historia*. Palma. n. p.

Salvà Tomàs, P. (1974) 'La parcelación, propiedad y utilización del suelo en el municipio de Andratx.' *BCOCIN*. 683.

Salvà Tomàs, P. (1983) 'Transformacion demografia y turismo: el cas de municipio de Calvia.'

Festes del Rei en Jaume, 1983, 9–13.

Salvà Tomàs, P. (1985) 'Turisme i canvi a l'espai de les Illes Balears.' *Treballs de la societat catalana de geografia.* 2, 29.

Salvà Tomàs, P. (1986) 'Aproximació al coneixement del paisatge de la muntanya de l'illa de Mallorca.' In: *Quinze ans del premis d'investigació. Ciutat de Palma 1970–84.* Palma. Ajuntament de Palma

Salvà Tomàs, P. (1987) 'Les conseqüències del desenvolupament del turisme sobres les activitats agràries a l'espai de les Illes Balears in Mallorca.' *Ara.* Fundació Emili Darder. Palma, 21–31.

Salvà Tomàs, P. (1990) 'El Turisme com una element impulsor del procés d'urbanització a Balears (1960–1989).' In: Fuster, M. A. (ed.) '30 anys de turisme a Balears.' *Estudis Baleàrics* 37–38, 63–70.

Salvà Tomàs, P. (1992) 'Els effects de la transició demografia illenca sobre el territorí: el marc de l'emigració a les illes Balears entre 1878 i 1955.' Papers to *Congrès Internacional d'estudis historics.* Palma, 405–411.

Salvà Tomàs, P. (1993) 'Changes and perspectives in agricultural land-use and their geoecological consequences for the mountain of Mallorca island.' *Pirineos.* 141–2: 92.

Salvà Tomàs, P. (ed.) (1995) *Atles de les Illes Balears.* Palma. Edicions Cort and Govern Balear.

Salvà Tomàs, P. (2000) 'El turisme de mañana.' In: Tugores Truyol, F. (Co-ordinator): *Welcome! Un siglo de turismo en las Islas Baleares.* Catalogue of exhibition. Fondació 'la Caixa'. Barcelona, 124–34.

Samuel, R. (1994) *Theatres of memory.* London. Verso.

Sanchez-Albornoz, N. (ed.) *Economic modernization of Spain, 1830–1930.* New York. New York University Press. Trans. Powers, K. and Sanudo, M.

Sand, G. (1855) *Un hiver en Majorque.* Trans. R. Graves. (1957). Valldemossa. n. p.

Sandez Leon, M. L. and Garcia Riaza, E. (2005) 'Romanization in the Balearic Islands.' In: Tugores Truyol, F. (Co-ordinator): *El mon róma a les illes Baleares.* Catalogue of exhibition. Barcelona. Fondació 'la Caixa'.

Sansó Barceló, S. (2009) *La indústria de les perles a Manacor (1902–2002).* Palma. Institut Balear d'Económia.

Sansó Barceló, S. (2009) *La Indústria de les perles a Manacor (1902–2002).* Llibres de l'Insitut Balear d'Economia 2. Palma. Institut Balear d'Economia.

Sansó Barceló, S. (2011) *Els fusters de Manacor.* Manacor. Associació empresarial de la fusta de Balears.

Santandreu Sureda, J. (1990) 'Turisme i marginació.' In: Anelyar Fuster, M. (ed.) (1990) '30 anys de turisme a Balears.' *Estudis Baleàrics,* 37–38. Palma. Govern Balear, 183–89.

Sbert Barceló, T. (2007) *Una evolución turística; historia de la playa de Palma 1900–2006.* S'Arenal. AETPP.

Schama, S. (1996) *Landscape and memory.* London. Fontana.

Seguí Aznar, M. (1999) 'L'arquitectura i l'urbanisme en el canvi de dècada (1920–30). El programa municipal de l'Assembla dels partits Autonomista i Regionalista.' In: Marimar Riutort, A. and Serra Busquets, S. (eds) 'Els anys vint a les Illes Balears.' *17 Jornades d'estudis històrics locals.* Palma. Institut d'estudis Baleàrics, 57–68.

Seguí Aznar, M. (2001) *La arquitectura del ocio en Baleares.* Lleonard Muntaner. Palma.

Seguí, J. (1998) *Les balears en venda; la desinversió immobiliària dels illencs.* Palma. Documenta Balear.

Segura, A. and Suau, J. (1986) 'The demographic history of Mallorca.' *Boletin Asociacion Demografia Historia.* 4 (1): 52–88.

Segura, A. and Suau, J. (1984) 'Estudi de demograf mallorquina: l'evolucie de la poblacion.' *Randa,* 6, 19–62.

Serra i Busquets, S. (2001) *Els elements de canvi a Mallorca del segle XX.* Palma. Cort.

Shelley, H. C. (1926) *Majorca.* London. Methuen.

Shubert, A. (1990) *A social history of modern Spain.* London. Routledge.

Siegfried, A. (1948) *The Mediterranean.* London. Jonathan Cape. English trans., Doris Hemming.

Simpson, James (1995) *Spanish agriculture: the long siesta, 1765–1965.* Cambridge. Cambridge University Press.

References

Smith, D. and Buffery, H. (2003) *The Book of Deeds of James I of Aragon: a translation of the medieval Catalan Llibre dels Fets.* Aldershot. Ashgate.

Socias Fuster, M. (1989) 'Los campos de golf en Balears: la nueva oferta complementaria.' In: *XI Congreso nacional de geografía; comunicaciones.* Vol 3. 403–10. Madrid. Asociación de geógrafos españoles.

Soler Gaya, R. (2004) *Cronica de los puertos baleares.* Palma. Edicions Documenta Balear.

Soto i Company, R. (1990) 'Repartiment and repartiments: l'ordinació d'un espai de colonició feudal a la Mallorca in de siegle XIII.' In: *De Al Analus a la sociedad feudal: los repartimentos bajo medieval.* Barcelona. CSIC 1–51.

Soto i Company, R. (1994) 'Porcio de Nuñio Sanç. Repartiment i repoblació de les terres del Sud Est de Mallorca.' *Afers* 18, 347–66.

Suau i Puig, J. (1991) El mon rural mallorquí, segles 18 and 19. Barcelona. El Curial.

Tabak, F. (2008) *The waning of the Mediterranean 1550–1870; a geohistorical approach.* Baltimore. Johns Hopkins University Press.

Taylor, C. (1983) *Village and farmstead; a history of rural settlement in England.* London. Philip.

Tello, E. (2006) 'La transformacion del territorio, antes y despues de 1950: un lugar de encuentro transdicipliniar para el estudio del paisaje.' *Areas,* 25, 5–11.

Thompson, F. M. L. (1968) *Chartered surveyors: the growth of a profession.* London. Routlege and Kegan Paul.

Tortella, G. (1987) 'Agriculture – a slow moving sector, 1830–1935.' In: Sanchez-Albornoz, N. (ed.) *Economic modernization of Spain, 1830–1930.* New York. NYUP. Trans. K. Powers and M. Sanudo.

Toubert, P. (1998) 'Incastellamento.' In: Barceló, M. and Toubert, P. (eds) *L'incastellamento: actes des rencontres de Gérone 22–27 Nov 1992 and de Rome 5–7 May 1994.* Rome. Boccard.

Tous Meliá, J. (2004) 'La evolución urbana de Palma, una visión icnográfica.' *Revista bibliográfica de geografía y ciences sociales* (Serie documental de *Geo Crítica*) IX, 518.

Tous Meliá, J. (2004) 'Palma a través de la cartografia (1596–1902).' *Revista bibliografica de geografia y ciences sociales* (Serie documental de *Geo Crítica*), IX, 515.

Tous Meliá, J. (2009) 'Palma a través de la cartografia (1596–1902).' Palma. Ajuntament de Palma.

Trias Mercant, S. (1996) *Les possessions mallorquines de l'Arxiduc.* Palma. Joan Muntaner.

Tugores Truyol, F. (Co-ordinator) (2000) *Welcome! Un siglo de turismo en las Islas Baleares.* Catalogue of exhibition. Fondació 'la Caixa'. Barcelona.

Tugores Truyol, F. (Co-ordinator) (2005) *El mon róma a les illes Baleares.* Catalogue of exhibition. Fondacio 'la Caixa'. Barcelona.

Turner, M., Gardner, R. H. and O'Neill, R. V. (2001) *Landscape ecology in theory and practice: pattern and process.* New York. Springer.

Umbert, J. A. (1991) 'El sector extractiva i industries derivades in Mallorca 1921–31.' In: Manera, C. and Petrus, J. M. (eds) *Del taller a la fàbrica.* Palma. Ajuntament de Palma.

Unwin, T. (1996) *Wine and the Vine; an historical geography of viticulture and the wine trade.* London. Routledge.

Valero i Martí, G. (ed.) (1989) *Elements de la societat pre-turistica mallorquina.* Palma. Govern Balear. Consellaria de Cultura, Educació I Esports.

Valero i Martí, G. (1992) *Itineries pel centre històric de Palma.* Col·lació Palma Ciutat educativa, no. 21. Palma. Ajuntament de Palma.

Valero i Martí, G. (1996) 'Notes sobres la transhumància a Mallorca: el camí de muntanya de la Torre de Llucmajor al Teix.' *Jornades d'estudis locals a Marratxí.* 1, 241–49.

Valero i Martí, G. (2011) *Guia del paisatge cultural de la Serra de Tramuntana.* Palma. Lleonard Muntaner.

Van Strydonck, M., Boudini, M. and Ervynck, A. (2002) 'Stable isotopes (^{13}C and ^{15}N) and diet; animal and human bone collagen from prehistoric sites on Mallorca, Menorca and Formentera (Balearic Islands, Spain).' In: Waldren, W. H. and Ensenyat, J. A. (eds) *World Islands in Prehistory: Insular Investigations. Vth Deia International Conference of Prehistory.* BAR International Series 1095. Oxford, 189–197.

Vaquer Bennàssar, O. (1978) *Aspectes socioeconòmics de Manacor al segle XVI.* Palma. Moll.

References

Vaquer Bennàssar, O. (1983) *Una societat de l'antic règim: Felanitx i Mallorca al segle XVI*. Felanitx. Ajuntament de Felanitx.

Varga Ponce, J. (1787) *Descripciones de las isles Pithiusa y Baleares*. Madrid. 1983 facimile with introduction by Isabel Moll i Blanes. Palma. Olañeta.

Vassberg, D. E. (1984) *Land and society in Golden Age Castile*. Cambridge. Cambridge University Press.

Vidal, C. (1994) 'La planta hotelera a la badia d'Alcúdia a la dècada dels anys seixanta.' *13 Jornades d'estudis històrics locals*. El desenvolupament turístic a la Mediterrània durant el segle XX. Palma. Institut d'Estudis Baleàrics, 247–54.

Vidal Ferrando, A. (1992) *La població i la propietat de la terra en la municipi de Santanyí (1868–1920)*. Monografies d'Història Local 5, Santanyí. Ajuntament de Santanyí.

Vidal Nicolau, A. (1991) 'Les activitats industrials a Llucmajor (1870–1936). De fariners a la importància del calcat.' In: Manera, C. and Petrus, J. M. (eds) *Del taller a la fàbrica*. Palma. Ajuntament de Palma.

Vilà Valentí, J. (2001) 'Informe sobre la tesis doctoral de Jean Bisson (1974).' *Territoris*, 3, 287–292.

Vinas, A. and R. (2004) *La conquête de Majorque; textes et documents*. Perpignan. Societe agricole, scientifique et litteraire des Pyrenne-Orientales. Vol CXI.

Vivot, T. (2006) *Les possessions de Mallorca. Vol 1*. Pollença. El Gall.

Vogiatzakis, I., Pungetti, G. and Mannion, A. (2008) *Mediterranean island landscapes: Natural and cultural approaches*. Berlin. Springer.

Waldren, J. (1996) *Insiders and outsiders; paradise and reality in Mallorca*. Providence RI. Berghahn Books.

Waldren, W. (2002) 'A case history: evidence of ancient animal, water and land management – Son Oleza Chacolithic Old Settlement, Valledemossa, Mallorca, Balearic Islands, Spain.' In: Waldren, W. H. and Ensenyat, J. A. (eds) *World Islands in Prehistory: Insular Investigations. Vth Deia International Conference of Prehistory*. BAR International Series 1095. Oxford, 301–11.

Waldren, W. (2002) 'Links in the chain: evidence of sustained prehistoric contact and cultural interaction between the Balearic Islands and Continental Europe.' In: Waldren, W. H. and Ensenyat, J. A. (eds) *World Islands in Prehistory: Insular Investigations. Vth Deia International Conference of Prehistory*. BAR International Series 1095. Oxford, 152–85.

Waldren, W. H. and Ensenyat. J. A. (eds) (2002) *World Islands in Prehistory: Insular Investigations. Vth Deia International Conference of Prehistory*. BAR International Series 1095. Oxford.

Walton, J. (2005) 'Paradise lost and found: tourists and expatriates in El Terreno, Palma de Mallorca from the 1920s to 1950s.' In: Walton, J. (ed.) *Histories of Tourism*. Clevedon. Channel View Publications, 179–94.

Waring, H. R. (1877–8) 'The drainage, irrigation and cultivation of the Albufera of Alcudia, Majorca: and application of the common reed as a material for paper.' *Minutes of the proceedings of the Institution of Civil Engineers*, 52 (pt II): 243–49.

Watson, A. M. (1974) 'The Arab agricultural revolution and its diffusion.' *Journal of Economic History* 34 (1): 8–35.

Whyte, I. D. (2002) *Landscape and history since 1500*. London. Reaktion Books.

Wiley, J. (2007) *Landscape*. London. Routledge.

Wood, C. W. (1888) *Letters from Majorca*. London. Richard Bentley and Son.

Index

Page numbers in *italic* denote illustrations. Page numbers in **bold** denote tables.